Multifaceted Protocols in Biotechnology, Volume 2

Azura Amid

Editor

Multifaceted Protocols in Biotechnology, Volume 2

 Springer

Editor
Azura Amid ⓘ
International Institute for Halal Research
and Training
International Islamic University Malaysia
Kuala Lumpur, Malaysia

ISBN 978-3-030-75581-2   ISBN 978-3-030-75579-9   (eBook)
https://doi.org/10.1007/978-3-030-75579-9

This Springer imprint is published by the registered company Springer Nature Switzerland AG
The registered company address is: Gewerbestrasse 11, 6330 Cham, Switzerland

Contents

1 Graviola Fruit Extraction by Ionic Liquid Microwave-Assisted
  Extraction (IL-MAE) 1
  Daddiouaissa Djabir and Azura Amid

2 Role of Ionic Liquids in the Enzyme Stabilization: A Case
  Study with *Trichoderma Ressie* Cellulase 23
  Amal A. M. Elgharbawy, Md Zahangir Alam,
  Muhammad Moniruzzaman, Nassereldeen Ahmad Kabbashi,
  and Parveen Jamal

3 Role of Ionic Liquids in the Processing of Lignocellulosic
  Biomass 35
  Amal A. M. Elgharbawy, Sharifah Shahira Syed Putra,
  Md Zahangir Alam, Muhammad Moniruzzaman,
  Nassereldeen Ahmad Kabbashi, and Parveen Jamal

4 Proliferation of Rat Amniotic Stem Cell (AFSC) on Modified
  Surface Microcarrier 63
  Nurhusna Samsudin, Yumi Zuhanis Has-Yun Hashim,
  Hamzah Mohd Salleh, and Azmir Ariffin

5 Application of Spectroscopic and Chromatographic Methods
  for the Analysis of Non-halal Meats in Food Products 75
  Abdul Rohman and Nurrulhidayah Ahmad Fadzillah

6 Identification of Potential Biomarkers of Porcine Gelatin 93
  Nur Azira Tukiran, Amin Ismail, Haizatul Hadirah Ghazali,
  and Nurul Azarima Mohd Ali

7 Gamma Ray Mutagenesis on Bacteria Isolated from Shrimp
  Farm Mud for Microbial Fuel Cell Enhancement
  and Degradation of Organic Waste 103
  Ayoub Ahmed Ali, Azura Amid, and Azhar Muhamad

8   Effect of Temperature on Antibacterial Activity and Fatty
     Acid Methyl Esters of Carica Papaya Seed Extract 117
     Muhamad Shirwan Abdullah Sani, Jamilah Bakar,
     Russly Abdul Rahman, and Faridah Abas

9   Solid-State Fermentation of Agro-Industrial Waste Using
     Heterofermentative Lactic Acid Bacteria 133
     Lina Oktaviani, Muhammad Yusuf Abduh, Dea Indriani Astuti,
     and Mia Rosmiati

10   Synthesis of Chitosan-Folic Acid Nanoparticles as a Drug
     Delivery System for Propolis Compounds 145
     Marselina Irasonia Tan and Adelina Khristiani Rahayu

Abbreviations

BLAST	The Basic Local Alignment Search Tool
ELISA	Enzyme-linked Immunosorbent Assay
LC-MS/MS	Liquid Chromatography-Tandem Mass Spectrometry
MS/MS	Tandem Mass Spectrometry
NCBI	National Center for Biotechnology Information
PMF	Peptide Mass Fingerprinting
SDS-PAGE	Sodium dodecyl sulphate-polyacrylamide gel electrophoresis

Chapter 1
Graviola Fruit Extraction by Ionic Liquid Microwave-Assisted Extraction (IL-MAE)

Daddiouaissa Djabir and Azura Amid (ID)

Abstract This chapter discusses the extraction method of Graviola fruit using ionic liquid microwave-assisted extraction (IL-MAE). Three parameters—irradiation power, solid/liquid (s/l) ratio, and extraction time—were investigated to obtain optimum extraction process resulting in a high yield and a low IC_{50} value for the anti-proliferation assay. The results showed that the extraction time of 1.74 min, the irradiation power of 300 W and the s/l ratio of 1:25 gave the highest extraction yield of 67.6%. Whereas the extraction time of 3 min instead of 2.98 min, the irradiation power of 700 W instead of 690 W, and the s/l ratio of 1:39 produced an average IC_{50} of 4.75 ± 0.36 µg/mL for MCF7 and 10.56 ± 2.04 µg/mL for HT29. The IL-MAE technique was found to be effective in the extraction of plant fruit as described in the case study.

Keywords Ionic liquid · Graviola · RSM · Microwave-assisted

1.1 Introduction

Ionic liquid (IL) is an organic liquid salt comprising of an organic cation matched with an organic or inorganic anion. ILs are generally perceived solvents owing to their phenomenal properties, such as poor power conductors, non-ionizing (e.g. non-polar), high viscosity, low combustibility, low vapor pressure, excellent thermal stability, wide liquid regions, and good solvating properties for polar and non-polar compounds (Huddleston et al., 2001).

D. Djabir
Kulliyyah of Engineering, International Islamic University Malaysia, Kuala Lumpur 53100, Malaysia

A. Amid (✉)
International Institute for Halal Research and Training (INHART), International Islamic University Malaysia, Kuala Lumpur 53100, Malaysia
e-mail: azuraamid@iium.edu.my

A. Amid (ed.), *Multifaceted Protocols in Biotechnology, Volume 2*,
https://doi.org/10.1007/978-3-030-75579-9_1

1.2   Ionic Liquid Microwave-Assisted Extraction (MAE)

The role of ILs is not limited to enhanced interactions between solute and solvent, which means increased solute solubility, but can be attributed to solvent-matrix interactions leading to changes in the permeability of the plant matrix. This is due to the disruption of the cell tissue and to the modification of the matrix permeability by the interaction of H bonding with the carbohydrates that form the cell walls. ILs could be considered as valuable solvent substituents for the extraction of value-added chemicals, especially when coupled with microwave-assisted extraction (MAE) and other methods known to disrupt targeted cell tissue that facilitate the whole process of extraction.

Higher yields and faster extraction of bio-compounds from biomass can be achieved through MAE processes. The pioneering work on the use of IL-MAE processes was first reported by Du and co-workers (2007), who demonstrated the effective utilization of aqueous IL solutions in MAE for the extraction of trans-resveratrol from *Rhizoma polygoni*.

1.2.1   Principle

Alupului and co-workers (2012) described the MAE mechanism involving three steps, starting from the separation of solutes from active matrix sites under increased pressure and temperature; then the distribution of solvent through sample matrix; finally, the release of solutes from the sample matrix to the solvent.

Summing up, almost all IL-based MAE processes reported to date are imidazolium-based ILs. The majority of the investigated ILs comprised of cations with short alkyl side chains length. Despite the lack of discussion on the mechanisms behind improved extraction yields, some authors have related the success of their extractions to the establishment of strong interactions, mainly hydrogen-bonding and $\pi-\pi$ interactions, between the ILs and the target bio-compounds. Various processing conditions, such as irradiation power, solid/liquid (s/l) ratio, and extraction time, chemical structure, and concentration of ILs, have been also addressed in previous studies. In general, high concentrations of ILs and low s/l ratios have promoted high extraction yields, whereas other parameters, such as irradiation power, have resulted in a lack of clear tendencies. Meanwhile, the use of tensioactive ILs promotes either an increase or a decrease in extraction yield, depending mostly on the hydrophilic-lipophilic ratio of the target bio-compound. There is therefore a need for more studies in this field, as many more conditions need to be explored, namely the size and type of IL aggregates (Ventura et al., 2017).

Table 1.1 Consumable used in this experiment

No.	Consumable	Usage
1	Whatman 3 mm filter paper	Filtration
2	Test tube	Sample collection

Table 1.2 Equipment used in this experiment

No.	Equipment	Usage
1	Microwave	Extraction
2	Freeze dryer	Drying sample

Table 1.3 Chemicals used in this experiment

No.	Chemicals	Usage
1	Ionic liquid solution ([C4MIM]Cl−, [C4MIM]BF4− and [C4MIM]PF6−)	Extraction
2	Graviola fruit	Extraction

1.2.2 Objective of Experiment

The objective of this procedure is to identify the most suitable condition for Graviola fruit extraction (GFE) using the IL-MAE method.

1.3 Materials

Tables 1.1, 1.2, and 1.3 show the materials used in this experiment.

1.4 Methodology

1.4.1 Statistical Optimization Set-Up

The use of statistical optimization experiments on various factors related to the production process is well documented. These techniques are necessary for the simultaneous analysis of various factors under study. Besides, they reduce the number of experiments involved, enhance data interpretation, and reduce the overall experimental time required. There are many software packages available for such data analysis. This study used Design-Expert v.7.0.8 as powerful software for determining the optimum extraction parameters of IL-MAE that can increase the extraction yield and the IC_{50} of IL-GFE toward cancer cell lines.

1.4.2 One-Factor-at-a-Time (OFAT) Experimental Design

The classical one-factor-at-a-time (OFAT) experimental design was employed to determine the possible optimum IL to maximize the yield of Graviola fruit extract. Previous literature has shown that several factors are affecting IL-MAE, such as extraction time, irradiation power, s/l ratio, and the type of IL used for extraction (Zhang et al., 2014). To increase the extraction efficiency of the three analytes, a series of the 1-butyl-3-methyl-imidazolium cation with various anions Cl−, BF4− and PF6− were compared, including [C4MIM]Cl−, [C4MIM]BF4−, [C4MIM]PF6− and deionized water, while other extraction parameters were kept constant at zero, i.e. extraction time (2 min), microwave irradiation power (450 W) and s/l ratio [1:20] (g/mL) (Zhang et al., 2014).

1.4.3 Optimization of Extraction Conditions

The optimization experiments were designed using Design Expert software v.7 (Stat-Ease) to achieve the maximum extraction yield of Graviola fruit extract and the minimum IC_{50} toward cancer cell lines. Based on the Face-Centered Central Composite Design (FCCCD) under Response Surface Methodology (RSM), three independent variables, namely extraction time [A], irradiation power [B], and s/l ratio [C], with three levels were designed to increase response yield and IC_{50}. The center point (0), the lowest (−1), and the highest (+1) values for all factors were selected based on the previous literature (Table 1.4) (Maran et al., 2013).

The experimental design obtained from the FCCCD consists of 12 triplicates of experiments or runs. The experimental design used in this work is presented in Table 1.5.

All the experiments were carried out in triplicates. The percentage of extraction yield and half-maximal inhibition IC_{50} of IL-GFE toward cancer cell lines were considered to be the response (Y). The evaluation of the analysis of variance (ANOVA) was achieved by conducting a statistical analysis of the model. The relationship between the responses (dependent factors) and the experimental levels of each variable under study was described, and the fitted polynomial equation was expressed in the form of contour and response surface plots.

Table 1.4 Experimental design and levels of independent process variables

Independent variables	Symbol	Levels of independent variables		
		−1	0	+1
Extraction time (min)	A	1	2	3
Irradiation power (W)	B	300	500	700
Solid-liquid ratio (gm/L)	C	(1:20)	(1:30)	(1:40)

Table 1.5 Experimental design for optimization of extraction parameters using RSM

Run	Process variables			Responses	
	Factor 1 [A]	Factor 2 [B]	Factor 3 [C]	Extraction yield (%)	Cytotoxicity IC 50 (µg/mL)
	Time (min)	Power (W)	Solid-liquid ratio (g/mL)		
1	1.00	300.00	20.00		
2	2.00	500.00	20.00		
3	3.00	300.00	40.00		
4	3.00	500.00	30.00		
5	2.00	500.00	20.00		
6	1.00	500.00	30.00		
7	2.00	500.00	40.00		
8	3.00	700.00	20.00		
9	3.00	500.00	30.00		
10	2.00	500.00	40.00		
11	2.00	300.00	30.00		
12	2.00	500.00	30.00		
13	2.00	500.00	30.00		
14	1.00	700.00	40.00		
15	3.00	300.00	40.00		
16	2.00	300.00	30.00		
17	2.00	500.00	40.00		
18	2.00	500.00	30.00		
19	2.00	700.00	30.00		
20	1.00	300.00	20.00		
21	2.00	500.00	30.00		
22	1.00	700.00	40.00		
23	3.00	500.00	30.00		
24	3.00	700.00	20.00		
25	1.00	300.00	20.00		
26	2.00	500.00	20.00		
27	2.00	500.00	30.00		
28	2.00	700.00	30.00		
29	2.00	300.00	30.00		
30	1.00	500.00	30.00		
31	2.00	500.00	30.00		
32	3.00	700.00	20.00		
33	3.00	300.00	40.00		

(continued)

Table 1.5 (continued)

Run	Process variables			Responses	
	Factor 1 [A]	Factor 2 [B]	Factor 3 [C]	Extraction yield (%)	Cytotoxicity IC 50 (μg/mL)
	Time (min)	Power (W)	Solid-liquid ratio (g/mL)		
34	1.00	500.00	30.00		
35	1.00	700.00	40.00		
36	2.00	700.00	30.00		

1.4.4 Validation of the Experimental Model

The optimum result predicted by the model was validated concerning all three independent variables. Another set of extractions selected as predicted by the model of the Design Expert software were used to validate the maximum yield and minimum IC_{50} under defined conditions.

1.5 Results and Discussion

1.5.1 Extraction of Graviola Fruit Extract

IL-MAE was applied on Graviola fruit to increase the extraction yield and cytotoxicity of the extracted compounds against selected cancer cell lines. The IL-MAE method, using the optimum extraction parameters, produced 66.6% extraction yield as shown in Fig. 1.4 (Sect. 1.6.8). This yield was high for solvent used and short extraction time.

This result indicates the efficiency of ILs by modifying the permeability of the plant matrix through the interaction of H bonding with the carbohydrates that form the cell walls (Bogdanov, 2014), whereas microwave heating induces cell-wall disruption and accelerates diffusion through membranes that facilitates targeted analyte release (Guo et al., 2017). On the other hand, previous studies reported lower extraction yields from Graviola fruit using conventional extraction methods such as Soxhlet and maceration, which usually take many hours for the same extraction yield (Melot et al., 2009; Ragasa et al., 2012). According to Lu et al. (2008), the extraction yield from these conventional extraction methods is not economically beneficial because it is time-consuming and solvent-consuming.

1.5.2   The Best Ionic Liquid Solution by OFAT

Performance of 1-butyl-3-methyl-imidazolium cation with various anions (Cl−, BF4− and PF6−) in the GFE process using the OFAT experimental design was evaluated to observe the effects of various anions and deionized water on the extraction yield. The structure of ILs has a great influence on their physicochemical properties, which may have an impact on the extraction efficiency of targeted analytes (Zhang et al., 2011). In the series of ILs studied in this work, the Cl− anion showed a higher extraction yield than others for the examined sample, as shown in Fig. 1.1.

It has been suggested that ILs with Cl− hydrophilic anion, could miscible in any proportion of water (Lu et al., 2008), thus yielding better extraction yields. A previous study conducted by Cláudio et al. (2013) reported that an aqueous solution of [C4MIM]Cl is the suitable extractive solvent of caffeine from *Paullinia cupana* seeds (guaraná) in which, under optimized extraction conditions (2.34 M [C4MIM]Cl, 70 °C, 30 min, 1:10 s/l ratio); the extraction efficiency increased by 50% compared to that obtained by Soxhlet extraction method using dichloromethane solvent. In another study, Jin and co-workers (2011) explored the ability of [C4MIM]Cl to improve the release of phenolic aldehyde paeonol from the roots of *Cynanchum paniculatum*. The authors reported that the IL-assisted extraction in the optimum extraction conditions (70 °C, 8 h, 1:7.3 s/l ratio) produced a higher extraction yield than that of the Soxhlet extraction technique.

On the other hand [C4MIM][BF4] gave a lower extraction yield of Graviola fruit extract. A similar study was carried out using 1.5 M [C4MIM][BF4] and 1 M [C6MIM][BF4] to extract phenolic alkaloids liensinine, neferine, and isoliensinine from the *Nelumbo nucifera* seeds (Lu et al., 2008). The authors showed that the optimized extraction conditions of IL-MAE have increased the extraction efficiency

Fig. 1.1 Effect of different ionic liquids on extraction yield

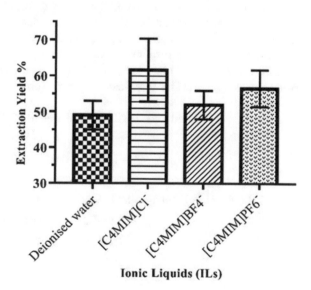

by 20–50% compared to conventional heat reflux extraction (HRE) and MAE with 80% methanol, for a significantly reduced extraction time from 2 h (HRE) to 1.5 min (IL-MAE). However, a strong dependency has been observed between the s/l ratio and extraction efficiency; the efficiency was found to have decreased when the s/l ratio increased from 1:10 to 1:20. The increase of viscosity by increasing the chain length of the imidazolium cation was proposed as an explanation of this phenomenon.

Moreover, [C4MIM][PF6]—relatively hydrophobic—is only sparingly water-soluble and therefore gave lower extraction yield. It is hypothesized that the extraction yield is affected by the degree of hydrophobicity (Morais et al., 2017). Previously, Zhai et al. (2009) used pure [C4MIM][PF6] as a water substitute coupled with MAE for the extraction of essential oils from two commonly used cooking spices, known as *Cuminum cyminum* (cumin) and *Illicium verum* (star anise). In this study, ILs are found to be able to absorb microwave energy more easily than water allowing the appropriate extraction temperature to be reached approximately three times faster. Thus, IL-MAE has considerably shortened extraction time (15 min) compared to conventional hydrodistillation (180 min) to completely extract essential oils. Also, they reported that the use of IL-MAE has reduced the oxidation and hydrolyzation of the essential oil constituents.

1.5.3 Optimization of the IL-MAE Parameters by RSM

The regression analysis used ANOVA. The FCCCD under RSM was employed to investigate the optimal parameters of the three variables (extraction time, irradiation power, and s/l ratio) in order to maximize the extraction yield and minimize IC_{50} of the crude IL-GFE on MCF7 and HT29 cancer cell lines. The experimental results and the predicted yield along with IC_{50} obtained from the regression equations are presented for each run in Table 1.6. The second-order polynomial equation was fitted to the data by multiple regression procedures. Each run has a unique combination of factor levels and the responses: extraction yield (%) (Y1) and IC_{50} (μg/mL) of IL-GFE on MCF7 (Y2) and HT29 (Y3) cells. The model equations obtained are listed as Eqs. (1.1–1.3) for three-factor systems (process conditions).

$$Y1 \ (\text{Yield}, \ \%) = 81.94 - 9.66A - 0.15B + 1.02C$$
$$+ \ 0.05AB + 0.48AC + 0.004BC$$
$$- 1.66A^2 - 0.04C^2 - 0.001ABC \tag{1.1}$$

$$Y2 \ (\text{MCF7} - \text{IC50}) = 50.7 - 98.62A - 0.2B + 0.27C + 0.25AB$$
$$+ \ 3.64AC + 0.01BC - 8.88A^2 - 0.0002B^2$$
$$- \ 0.09C^2 - 0.007ABC \tag{1.2}$$

Table 1.6 FCCCD experimental design and result of the responses

Run	Extraction condition			Y_1: yield (%)		Y_2: IC$_{50}$ of IL-MAE on MCF7 cells		Y_3: IC$_{50}$ of IL-MAE on HT29	
	A: extraction time (min)	B: irradiation power (W)	C: solid—liquid ratio (1 g/mL)	Actual	Predicted	Actual	Predicted	Actual	Predicted
1	3	700	20	56.66±0.58	57.23	15.40±1.25	15.65	14.95±2.44	14.76
2	3	300	40	54.66±0.58	55.05	14.42±1.71	13.88	39.09±8.92	39.01
3	1	700	40	52.00±1.00	52.45	19.16±0.68	19.55	24.66±1.22	24.51
4	1	300	20	65.66±1.15	66.03	10.21±1.71	10.23	11.69±1.27	11.66
5	1	500	30	58.66±1.15	59.11	22.37±0.59	23.03	29.8±0.59	29.72
6	3	500	30	59.00±1.00	59.36	13.66±0.78	13.31	12.17±0.48	12.01
7	2	300	30	66.33±1.53	66.38	14.70±0.31	14.45	28.87±1.83	27.95
8	2	700	30	55.33±1.53	55.42	22.54±0.08	22.85	11.85±0.88	10.79
9	2	500	20	59.00±1.00	59.48	8.12±0.33	8.22	12.99±2.06	12.89
10	2	500	40	53.00±1.00	53.32	28.30±1.41	28.00	12.46±1.32	12.32
11	2	500	30	60.66±0.58	60.90	26.81±0.77	27.05	18.46±1.65	19.37
12	2	500	30	61.00±1.00	60.90	27.50±1.8	27.05	18.79±1.39	19.37

$$Y3\ (HT29 - IC50) = -51.4 - 36.6A + 0.0002B + 4.3C$$
$$+ 0.08AB + 1.52AC + 0.001BC + 1.49A^2$$
$$- 0.07C^2 - 0.004ABC \qquad (1.3)$$

where Y is the predicted responses; Y1 is the extraction yield of Graviola fruit; Y2 is the IC50 of IL-MAE on MCF7 cell; Y3 is the IC50 of IL-MAE on HT29 cell; whereas A, B and C are the coded values for extraction time, irradiation power, and s/l ratio, respectively.

Based on the results of the experimental design; the yield of Graviola fruit extract ranged between 52 and 66.33%. In which the yield (%) was maximum at a low value of microwave irradiation power (B), mid-value of extraction time (A), and mid-value of s/l ratio (C). The yield was minimum at a low value of extraction time (A), and a high value of microwave irradiation power (B), and a high value of s/l ratio (C). The values of B and C have a significant effect on extraction efficiency (Table 1.6).

On the other hand, IC_{50} values of IL-GFE ranged between 8.12 and 28.3 μg/mL when MCF7 cells were treated under different IL-MAE conditions, and IC_{50} values of IL-GFE ranged from 10.79 to 39.01 μg/mL when HT29 cells were treated with different IL-MAE conditions. The IC_{50} value of IL-GFE on MCF7 cell was the lowest when the lowest value of the s/l ratio (C) and mid values of extraction time (A) and irradiation power (B) were applied; whereas IC_{50} was maximal at the highest value of s/l ratio (C) and mid values of extraction time (A) and irradiation power (B). Besides, the IC_{50} value of IL-GFE on HT29 was minimal when the highest value of irradiation power (B) and mid values of extraction time (A) and s/l ratio (C); whereas the IC50 value was maximum when treated on HT29 at the highest values of extraction time (A) and s/l ratio (C), and lowest value of irradiation power (B).

1.6 Discussion

1.6.1 Analysis of Variance

The F-test and p-value evaluated the statistical significance of the regression model. ANOVA for the fitted polynomial models of GFE yield is presented in Table 4.2, whereas the IC50 for MCF7 and HT29 are presented, respectively in Tables 1.9 and 1.10. The Quadratic model was applied in this case because of its high coefficient of determination (R^2) and its ability to enhance model terms.

1.6.2 Analysis of Variance of the Extraction Yield

The level of significance and adequacy of the generated model was assessed by ANOVA, considering the p-value. Generally, the model terms with P-value less than 0.05 are significant. The p-value < 0.0001 and F-value of 52.4 indicate that the model term is highly significant (Table 1.7).

Besides, the p-value is used to check the significance of the individual coefficient and their corresponding interactions between the variables. Additionally, the response showed that the linear coefficients of power (B) and s/l ratio (C), interaction terms—AC (contact time and s/l ratio), BC (contact power and s/l ratio), ABC (contact time, power and s/l ratio)—and the quadratic coefficients—A2 and C2—were also significant ($p < 0.05$) and considerably affected the extraction yield. The very small p-values ($p < 0.05$) indicated that the extraction power and s/l ratio were correlated with the extraction yield. The adequate precision of the model can be verified using the lack of fit. The result of ANOVA showed that LOF is not significant ($P > 0.05$) with a p-value of 1.00 relative to the pure error of extraction yield.

Table 1.7 ANOVA for yield (%) fitted quadratic model of extraction conditions

Source	Sum of squares	Degree of freedom	Mean squares	F-value	P-value
Model	658.28	9	73.14	52.4	<0.0001*
A-extraction time	0.17	1	0.17	0.12	0.7325
B-iradiation power	187.04	1	187.04	134	<0.0001**
C-solide-liquid ratio	54	1	54	38.69	<0.0001**
AB	3.06	1	3.06	2.19	0.1506
AC	27.56	1	27.56	19.75	0.0001**
BC	11.67	1	11.67	8.36	0.0076
A^2	11.11	1	11.11	7.96	0.009
C^2	81	1	81	58.03	<0.0001**
ABC	17.5	1	17.5	12.54	0.0015*
Residual	36.29	26	1.4	–	–
Lack of fit	0.75	1	0.75	0.53	0.4744^NS
Pure error	35.54	25	1.42	–	–
Cor total	694.58	35	–	–	–

*$P < 0.05$ means the model terms are significant
**$P < 0.01$ means the model terms are highly significant
NS: Not significant

Table 1.8 The analysis of
the model fitting

Elements	Values	Elements	Values
Standard deviation (SD)	1.02	R^2 (R-Squared)	0.9617
Mean	58.50	Adjusted R^2	0.9485
C.V.	1.74	Predicted R^2	0.9244
PRESS	52.96	Adeq precision	26.770

1.6.3 Test for Significance of the Regression

Following the response surface regression processes, R^2 was calculated to confirm the significance of the model and to compare the actual and adjusted R^2. The closer the actual R^2 to the adjusted R^2, the more significant the model terms are. The actual and adjusted R^2 are shown in Table 1.7. If R^2 is closer to 1, the correlation between the observed and the predicted values are better. The higher values of R^2 (0.96) and the closest to the adjusted R^2 (0.948) for extraction yield also indicated the efficacy of the model, which suggests that the model equation could account for 96.17 and 94.85% variation. Thus, when the R^2 value is closer to 1, the model is better fitted (Table 1.8).

Adequate precision compares the range of the predicted values with the average prediction errors at the design points. Additionally, the signal-to-noise ratio was measured in which values of more than 4 are considered desirable (Ma et al., 2011). Adequate precision of 26.77 for extraction yield indicates an appropriate signal and the generated model can be used to navigate the design space. The coefficient of variation (CV) is the ratio of the standard error and the lower the CV value the reliability of the experiment achieved. The model is commonly considered to be reasonably reproducible if its CV is less than 10% (Ma et al., 2011). A low CV value of 1.74 for extraction yield suggested an excellent precision and reliability of the experiment.

1.6.4 The Interaction Response Effects

The three-dimensional (3D) response surface plots demonstrated the interaction effects of the independent variables on the dependent one. Also, surface graphs were used to select the optimum conditions to maximize the response. Figure 1.2 represents the interactions between the experimental levels of two tested variables and their impact on the response while the third variable was kept constant at zero. Various shapes of the contour plots indicate various interactions between different variables. Elliptical contours appeared when there was a typical interaction between variables (Zhang et al., 2011).

Figure 1.2a illustrates the impact of extraction time and irradiation power and their interaction on the extraction yield when fixing the s/l ratio at 1:30. The yield

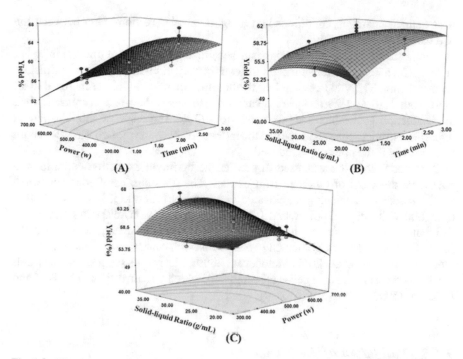

Fig. 1.2 3D response surface plots show the extraction parameters effect of IL-MAE on the yield of Graviola fruit. **a** The effect of extraction time vs. irradiation power on the yield; **b** The effect of extraction time vs. solid–liquid ratio on the yield; **c** The effect of irradiation power vs solid–liquid ratio on the yield

of Graviola fruit extract increased with the increase of extraction time from 1 to 2 min and then decreased. Microwave irradiation takes a certain amount of time to cause cell wall disruption and release the target component. However, the prolonged heating time led to a decrease in extraction yield because of thermal degradation and polymerization of the Graviola fruit sample (Guo et al., 2017). On the other hand, the increase in irradiation power from 300 to 700 W decreased the extraction yield. The irradiation power influences interactions and equilibrium rates and controls the analyte partitions between the sample and extraction phase (Ma et al., 2010). The elliptical contour shape of the response surface curve suggests a slight interaction between the two variables.

Figure 1.2b showed the response surface graph of extraction time and s/l ratio and their interactions on the extraction yield when fixing the power at 500 W. The yield of Graviola fruit increased with an increase in extraction time from 1 to 2 min and s/l ratio from 1:20 to 1:25 and then the yield decreased. Smaller solvent volumes would make the target extraction incomplete, while a large quantity of solvent could make the extraction process more complicated in addition to unnecessary waste (Ma et al., 2010). The 3D response surface plots show an elliptical contour shape indicating well defined operating conditions and a significant interaction effect between the two variables.

Figure 1.2c illustrated the response surface graph of the effect of irradiation power and s/l ratio and their interactions on the extraction yield while fixing the extraction time at 2 min. The yield of Graviola fruit extract increased with an increase in s/l ratio from 1:20 to 1:25 and then decreased. However, the yield decreased when the irradiation power increased from 300 to 700 W. The response plot showed an imperfect elliptical contour shape and the interaction between the two variables could be improved if the power was reduced.

The model equation was used to predict the optimum extraction conditions to maximize the yield of Graviola fruit extract. The optimum conditions suggested by the model for achieving an elevated yield are as follows: 1.74 min extraction time, 300 W irradiation power, and 1:25 s/l ratio to achieve 67.6% extraction yield. Compared to other studies, the extraction yield obtained in this study using IL-MAE was higher than that obtained in earlier research using conventional extraction methods (Bonneau et al., 2017; Melot et al., 2009). The results suggest that the MAE mechanism was based on a cell-wall explosion, which was investigated by Paré and Belanger (1993).

1.6.5   Validation of the Model

In order to validate the developed model and verify the optimum results, another set of extractions was carried out. Due to the limitation of the instruments, the chosen optimum parameters were 2 min extraction time instead of 1.74 min, 300 W irradiation power, and 1:25 s/l ratio (g/mL). The results of validation experiments produced an average yield of $66.6 \pm 2.52\%$. These values are close to the predicted value of 67.64% and indicate the adequacy of the optimization model.

1.6.6   Analysis of Variance of the IC$_{50}$

The F-test and p-value evaluated the statistical significance of the regression model. The ANOVA for the fitted polynomial models of the IC$_{50}$ of IL-GFE on MCF7 and HT29, respectively, are shown in Tables 1.9 and 1.10. The polynomial model was chosen because of its high R^2 and its ability to improve the model terms.

The adequacy and level of significance of the generated model were assessed through ANOVA by considering the p-value. Model F-value of 128.33 and p-value of < 0.0001 for MCF7, and F-value of 35.43 and p-value < 0.0001 for HT29 indicate that the model is highly significant (Tables 1.9 and 1.10). The p-value is used to check for the significance of the individual coefficient and their corresponding interactions between the variables. Additionally, the response revealed that linear coefficients, all interaction terms, and quadratic coefficients were significant ($p < 0.05$) and had

Table 1.9 ANOVA for the IC_{50} of IL-GFE on MCF7 cells fitted quadratic model of extraction conditions

Source	Sum of squares	Degree of freedom	Mean square	F-value	P-value
Model	1566.09	10	156.61	128.33	<0.0001**
A-TIME	127.05	1	127.05	104.11	<0.0001**
B-POWER	92.12	1	92.12	75.49	<0.0001**
C-RATIO	611.05	1	611.05	500.72	<0.0001**
AB	262.39	1	262.39	215.01	<0.0001**
AC	8.25	1	8.25	6.76	0.0154*
BC	88.89	1	88.89	72.84	<0.0001**
A^2	236.74	1	236.74	194.00	<0.0001**
B^2	218.28	1	218.28	178.87	<0.0001**
C^2	239.95	1	239.95	196.63	<0.0001**
ABC	156.78	1	156.78	128.47	<0.0001**
Pure Error	30.51	25	1.22	–	–
Cor Total	1596.60	35	–	–	–

*$P<0.05$ means the model terms are significant
**$P<0.01$ means the model terms are highly significant

Table 1.10 ANOVA for the IC_{50} of IL-GFE on HT29 cells fitted quadratic model of extraction conditions

Source	Sum of squares	Degree of freedom	Mean square	F-value	P-value
Model	2666.16	9	296.24	35.43	<0.0001**
A-TIME	466.40	1	466.40	55.78	<0.0001**
B-POWER	435.03	1	435.03	52.03	<0.0001**
C-RATIO	0.43	1	0.43	0.051	0.8230
AB	364.36	1	364.36	43.58	<0.0001**
AC	130.91	1	130.91	15.66	0.0005**
BC	701.28	1	701.28	83.88	<0.0001**
A^2	8.94	1	8.94	1.07	0.3106
C^2	183.06	1	183.06	21.90	<0.0001**
ABC	140.31	1	140.31	16.78	0.0004**
Residual	217.38	26	8.36	–	–
Lack of fit	9.01	1	9.01	1.08	0.3083
Pure error	208.37	25	8.33	–	–
Cor total	2883.54	35	–	–	–

*$P<0.05$ means the model terms are significant
**$P<0.01$ means the model terms are highly significant

Table 1.11 The analysis of the model fitting

Statistical analysis	MCF7	HT29
R-Squared	0.98	0.92
Adjusted R-Squared	0.97	0.90
Coefficient of variation (CV)	5.93	14.71
Adequate Precision	33.05	18.75

remarkable effects on the IC_{50} of MCF7 cell lines. Also, the response revealed that the linear coefficients A (time) and B (power), all interaction terms, and the quadratic coefficient C (s/l ratio) were significant ($p < 0.05$) and had remarkable effects on the IC_{50} of HT29 cell lines. The very small p-values ($p < 0.05$) indicated that the extraction time, power, and s/l ratio were significantly correlated with the IC_{50} of MCF7 cell lines. The result of ANOVA for HT29 showed that the lack of fit is not significant ($p > 0.05$) with the p-value of 0.3083, which is relative to the pure error for the IC_{50} of the cell.

1.6.7 Test for Significance of the Regression

Following the response surface regression processes, R^2 was calculated to reaffirm the significance of the model and to compare the actual and adjusted R^2. The actual and adjusted R^2 are shown in Table 1.11. The higher values of R^2 (0.98) nearest to the adjusted R^2 (0.97) for IC_{50} of MCF7 and the value of R^2 (0.92) nearest to the adjusted R^2 (0.90) for IC_{50} of HT29 also indicated the efficacy of the model, suggesting that the model equation could account for 98.09 and 97.32% variation for MCF7 and 92.46 and 90.85% variation for HT29. Thus, when the R^2 value is closer to 1.0, the model is better fitted.

Adequate precision compares the range of the predicted values with the average prediction errors at the design points. Additionally, it measures the signal-to-noise ratio where values greater than four are considered optimal. Adequate precision of 33.05 for MCF7 and 18.75 for HT29 indicates an adequate signal and the generated model can be used to navigate the design space. CV is the ratio of the standard error and the lower the CV value, the higher the reliability of the experiment achieved. The model is commonly considered to be reasonably reproducible if its CV is less than 10% (Zhang et al., 2014). A relatively low CV value of 5.93 for MCF7 and 14.71 for HT29 suggested a good precision and reliability of the experiment.

1.6.8 The Interaction Response Effects

The 3D response surface plots were drawn to demonstrate the interaction effects of the independent variables on the dependent variable. Also, surface graphs are used

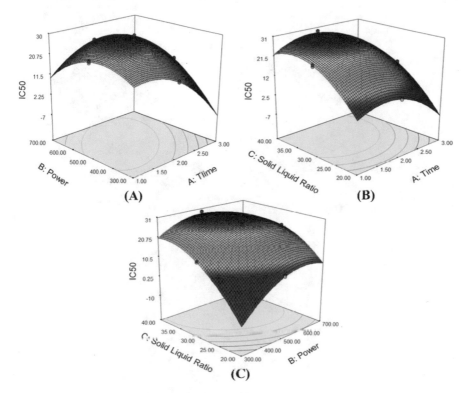

Fig. 1.3 3D response surface plots show the extraction parameters (IL-MAE) on the IC_{50} of MCF7 cell line. **a** The effect of extraction time vs irradiation power on the IC_{50} of MCF7; **b** The effect of extraction time vs solid–liquid ratio on the IC_{50} of MCF7; **c** The effect of irradiation power vs solid–liquid ratio on the IC_{50} of MCF7

to select the optimum conditions to minimize the response (IC_{50}). Figures 1.3 and 1.4 represent the interactions between the experimental levels of the two measured variables and their effect on the response while the third variable was kept constant at zero. Elliptical contours were obtained when there was a perfect interaction between the independent variable (Zhang et al., 2011).

Figures 1.3a and 1.4a illustrated the effect of extraction time and irradiation power and their interaction on IC_{50} while the s/l ratio (1:30) remained constant. The IC_{50} of IL-GFE on MCF7 cells decreased with the increase of time from 1 to 3 min and increased with the increase of power from 300 to 500 W then decreased. On the other hand, the IC_{50} of IL-GFE on HT29 cells decreased with an increase of extraction time from 1 to 3 min and an increase of irradiation power from 300 to 700 W. The elliptical contour shape of the response surface curve showed a slight interaction between these variables.

Figures 1.3b and 1.4b showed the response surface graph of extraction time and s/l ratio and their interactions with the IC_{50} when power was kept constant at 500 W. The IC_{50} of IL-GFE on MCF7 cells increased with an increase of time from 1 to

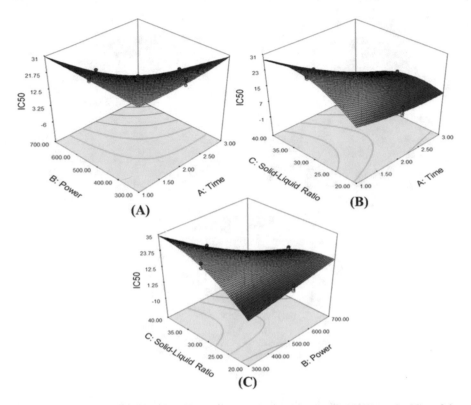

Fig. 1.4 3D response surface plots show the extraction parameters (IL-MAE) on the IC_{50} of the HT29 cell line. **a** The effect of extraction time vs irradiation power on the IC_{50} of HT29; **b** The effect of extraction time vs solid–liquid ratio on the IC_{50} of HT29; **c** The effect of irradiation power vs solid–liquid ratio on the IC_{50} of HT29

2 min and the s/l ratio increased from 1:20 to 1:35, then IC_{50} reduced. On the other hand, the IC_{50} of IL-GFE on HT29 cells increased with an increase of time (1–2 min) and s/l ratio (1:20–1:30), then IC_{50} reduced. The 3D response surface plots of MCF7 cells showed an elliptical contour shape indicating well defined operating conditions and a significant interaction effect between the two variables.

Figures 1.3c and 1.4c illustrated the response surface graph of the effect of irradiation power and s/l ratio and their interactions on the IC_{50} when extraction time was kept constant at 2 min. The IC_{50} of IL-GFE on MCF7 and HT29 increased with the increase of power (300–500 W) and s/l ratio (1:20–1:30), and then IC_{50} decreased. The response plot of HT29 showed an imperfect elliptical contour shape and the interaction between the two variables can be improved if the power is reduced.

The model equation was used to predict optimum extraction conditions. The optimum conditions suggested by the model to obtain the lowest IC_{50} are as follows: extraction time of 2.98 min, an irradiation power of 690 W and s/l ratio of 1:39; IC_{50} of 3.41 μg/mL for MCF7 and 6.61 μg/mL for HT29 was achieved under optimum

Table 1.12 Results from the validation of the model

	Extraction time (min)	Irradiation power (W)	Solid-liquid ratio (g/ml)	MCF7 IC_{50} (μg/mL)	HT29 IC_{50} (μg/mL)
Predicted	2.98	690	(1:39)	3.41	6.61
Actual	3	700	(1:39)	4.75 ± 0.36	10.56 ± 2.04

conditions. These results confirm the previous study on the anti-proliferative activity of isolated compounds from Graviola fruit against human prostate cancer PC-3 cells (Sun et al., 2016).

1.6.9 Validation of the Model

To validate the developed model and verify the optimum results, other sets of cyto-toxicity assays were carried out. Due to the instrumental limitation, the optimum extraction conditions were set to the following: extraction time of 3 min instead of 2.98 min, irradiation power of 700 W instead of 690 W, and s/l ratio of 1:39. The results from validation experiments yielded an average IC_{50} of 4.75 ± 0.36 μg/mL for MCF7 and 10.56 ± 2.04 μg/mL for HT29. These values are close to the predicted value indicating the adequacy of the optimization model (Table 1.12).

1.7 Conclusion

The IL-MAE technique presented was effective in extracting plant fruit as described in the case study.

Acknowledgements The author would like to acknowledge the International Islamic University Malaysia for awarding fund via Publication RIGS Grant (Grant No. P-RIGS18-065-0065).

References

Alupului, A., Calinescu, I., & Lavric, V. (2012). Microwave extraction of active principles from medicinal plants. *UPB Science Bulletin, Series B: Chemistry and Materials Science, 74*(2), 129–142.

Bogdanov, M. G. (2014). Ionic liquids as alternative solvents for extraction of natural products. In F. Chemat & M. A. Vian (Eds.), *Alternative solvents for natural products extraction.* Springer.

Bonneau, N., Baloul, L., Bajin ba Ndob, I., Sénéjoux, F., & Champy, P. (2017). The fruit of Annona squamosa L. as a source of environmental neurotoxins: From quantification of squamocin to annotation of Annonaceous acetogenins by LC–MS/MS analysis. *Food Chemistry, 226,* 32–40. https://doi.org/10.1016/j.foodchem.2017.01.042.

Cláudio, A. F. M., Ferreira, A. M., Freire, M. G., & Coutinho, J. A. (2013). Enhanced extraction of caffeine from guarana seeds using aqueous solutions of ionic liquids. *Green Chemistry, 15*(7), 2002–2010.

Du, F. Y., Xiao, X. H., & Li, G. K. (2007). Application of ionic liquids in the microwave-assisted extraction of trans-resveratrol from Rhizma Polygoni Cuspidati. *Journal of Chromatography A, 1140*(1–2), 56–62. https://doi.org/10.1016/j.chroma.2006.11.049.

Guo, J., Fan, Y., Zhang, W., Wu, H., Du, L., & Chang, Y. (2017). Extraction of gingerols and shogaols from ginger (Zingiber officinale Roscoe) through microwave technique using ionic liquids. *Journal of Food Composition and Analysis, 62,* 35–42. https://doi.org/10.1016/j.jfca. 2017.04.014.

Huddleston, J. G., Visser, A. E., Reichert, W. M., Willauer, H. D., Broker, G. A., & Rogers, R. D. (2001). Characterization and comparison of hydrophilic and hydrophobic room temperature ionic liquids incorporating the imidazolium cation. *Green Chemistry, 3*(4), 156–164. https://doi. org/10.1039/b103275p.

Lu, Y., Ma, W., Hu, R., Dai, X., & Pan, Y. (2008). Ionic liquid-based microwave-assisted extraction of phenolic alkaloids from the medicinal plant Nelumbo nucifera Gaertn. *Journal of Chromatography A, 1208*(1–2), 42–46. https://doi.org/10.1016/j.chroma.2008.08.070.

Jin, R., Fan, L., & An, X. (2011). Ionic liquid-assisted extraction of paeonol from cynanchum paniculatum. *Chromatographia, 73*(7–8), 787–792. https://doi.org/10.1007/s10337-010-1865-6.

Ma, C. H., Liu, T. T., Yang, L., Zu, Y. G., Chen, X., Zhang, L., Zhang, Y., Zhao, C. (2011). Ionic liquid-based microwave-assisted extraction of essential oil and biphenyl cyclooctene lignans from Schisandra chinensis Baill fruits. *Journal of Chromatography A, 1218*(48), 8573–8580. https:// doi.org/10.1016/j.chroma.2011.09.075.

Ma, W., Lu, Y., Hu, R., Chen, J., Zhang, Z., & Pan, Y. (2010). Application of ionic liquids based microwave-assisted extraction of three alkaloids N-nornuciferine, O-nornuciferine, and nuciferine from lotus leaf. *Talanta, 80*(3), 1292–1297. https://doi.org/10.1016/j.talanta.2009.09.027.

Maran, J. P., Sivakumar, V., Thirugnanasambandham, K., & Sridhar, R. (2013). Optimization of microwave assisted extraction of pectin from orange peel. *Carbohydrate Polymers, 97*(2), 703–709. https://doi.org/10.1016/j.carbpol.2013.05.052.

Melot, A., Fall, D., Gleye, C., & Champy, P. (2009). Apolar Annonaceous acetogenins from the fruit pulp of Annona muricata. *Molecules, 14*(11), 4387–4395. https://doi.org/10.3390/molecu les14114387.

Morais, T. R., Coutinho, A. P. R., Camilo, F. F., Martins, T. S., Sartorelli, P., Massaoka, M. H., Figueiredo, C. R., & Lago, J. H. G. (2017). Application of an ionic liquid in the microwave assisted extraction of cytotoxic metabolites from fruits of Schinus terebinthifolius Raddi (Anacardiaceae). *Journal of the Brazilian Chemical Society, 28*(3), 492–497. https://doi.org/10.21577/0103-5053. 20160215.

Paré, J. R. J., & Belanger, J. M. R. (1993). Microwave-assisted process (MAPTM): Applications to the extraction of natural products. In *Proceedings of the 28th Microwave Power Symposium, (MPSIMPI' 39)* (pp. 62–67).

Ragasa, C. Y., Soriano, G., Torres, O. B., Don, M.-J., & Shen, C.-C. (2012). Acetogenins from Annona muricata. *Pharmacognosy Journal, 4*(32), 32–37. https://doi.org/10.5530/pj.2012.32.7.

Sun, S., Liu, J., Zhou, N., Zhu, W., Dou, Q. P., & Zhou, K. (2016). Isolation of three new annonaceous acetogenins from Graviola fruit (Annona muricata) and their anti-proliferation on human prostate cancer cell PC-3. *Bioorganic & Medicinal Chemistry Letters, 26*(17), 4283–4285. https://doi.org/ 10.1016/j.bmcl.2015.06.038.

Ventura, S. P. M., e Silva, F. A., Quental, M. V., Mondal, D., Freire, M. G., & Coutinho, J. A. P. (2017). Ionic-liquid-mediated extraction and separation processes for bioactive compounds: Past, present, and future trends. *Chemical Reviews, 117*(10), 6984–7052. https://doi.org/10.1021/acs. chemrev.6b00550.

Zhai, Y., Sun, S., Wang, Z., Cheng, J., Sun, Y., Wang, L., Zhang, Y., Zhang, H., & Yu, A. (2009). Microwave extraction of essential oils from dried fruits of Illicium verum Hook. f. and Cuminum cyminum L. using ionic liquid as the microwave absorption medium. *Journal of Separation Science, 32*(20), 3544–3549. https://doi.org/10.1002/jssc.200910204.

Zhang, J., Jia, S., Liu, Y., Wu, S., & Ran, J. (2011). Optimization of enzyme-assisted extraction of the Lycium barbarum polysaccharides using response surface methodology. *Carbohydrate Polymers, 86*(2), 1089–1092. https://doi.org/10.1016/j.carbpol.2011.06.027.

Zhang, Y., Liu, Z., Li, Y., & Chi, R. (2014). Optimization of ionic liquid-based microwave-assisted extraction of isoflavones from Radix puerariae by response surface methodology. *Separation and Purification Technology, 129,* 71–79. https://doi.org/10.1016/j.seppur.2014.03.022.

Chapter 2
Role of Ionic Liquids in the Enzyme Stabilization: A Case Study with *Trichoderma Ressie* Cellulase

Amal A. M. Elgharbawyⓘ**, Md Zahangir Alam, Muhammad Moniruzzaman, Nassereldeen Ahmad Kabbashi, and Parveen Jamal**

Abstract This chapter discusses the stabilization of *Trichoderma ressie* cellulase (*Tri-Cel*) in ionic liquids (ILs) to enable the *in situ* hydrolysis of cellulosic and ligno-cellulosic substances. It is well recognized that enzymes tend to lose their activities in ILs, but several methods have been used to increase or minimize the loss of activity in ILs. In this study, cellulase was therefore tested in several ILs. This approach opens an insight for further studies to discover more about the effects of ILs on cellulase and their interactions in the aqueous system. It can also offer successful manufacturing and processing of different biomass biofuels.

Keywords Ionic liquid · Lignocellulose · Cellulase · Hydrolysis · Stability · Activity

2.1 Introduction

ILs may be described as organic liquid salts at room temperature and melt at or below 100 °C. They constitute a carbonic chain that produces a cation that is ioni-cally linked to an anion; thus, a wide variety of ILs may therefore be synthesized. In addition, ILs have customizable features, including thermal stability, miscibility and polarity, which are of significant advantages over traditional organic, non-reusable toxic and volatile solvents. (De Souza Mesquita et al., 2019). ILs have many desir-able characteristics, such as enzyme stabilization. Due to their merit, ILs are good media for various reactions (Elgharbawy et al., 2016; Fu et al., 2010). ILs can be

A. A. M. Elgharbawy (✉)
International Institute for Halal Research and Training (INHART), IIUM, Jalan Gombak, Malaysia
e-mail: amalgh@iium.edu.my

M. Z. Alam · N. A. Kabbashi · P. Jamal
Department of Biotechnology Engineering, Kulliyyah of Engineering, International Islamic University Malaysia, Kuala Lumpur 50728, Malaysia

M. Moniruzzaman
Chemical Engineering Department, Universiti Teknologi PETRONAS, Seri Iskandar, Malaysia

© The Author(s), under exclusive license to Springer Nature Switzerland AG 2021
A. Amid (ed.), *Multifaceted Protocols in Biotechnology, Volume 2*,
https://doi.org/10.1007/978-3-030-75579-9_2

applied as immobilization and coating agents for enzymes for diverse applications (Moniruzzaman et al., 2015). Enzymes may be activated or stabilized by ILs.

2.2 The Role of ILs in Enzyme-Catalyzed Hydrolysis

Enzyme-catalyzed hydrolysis of IL pretreated substrates involving cellulose transformation from IL solution for enzymatic hydrolysis can be demonstrated in two main pathways (Tan et al., 2011; Zhao et al., 2009). The primary pathway includes a multi-stage process where the biomass is pretreated, washed and then hydrolyzed to the desired product. The secondary pathway is considered a single-step method in which hydrolysis is performed in aqueous IL and cellulase enzymes (Gunny et al., 2014). Multiple ILs have shown impressive outcomes in structural modification of lignocellulose and removal of lignin, including choline acetate [Ch][Ac] (Asakawa et al., 2015). This demonstrates that ILs can be adapted for reliability with certain enzymes (Elgharbawy et al., 2016; Ibrahim et al., 2015).

2.2.1 Principle

Wang and co-workers (Wang et al., 2011a) reported that when analyzed in 1-ethyl-3-methylimidazolium acetate [EMIM][Ac] (15%), certain cellulases were maintained along the process of saccharification. The biomass of yellow poplar and [EMIM][Ac] with the percentage of 10–20%, was used for enzymatic hydrolysis (Shi et al., 2013). In addition, in ionic liquid-enzyme (IL-E) compatible systems, several studies have documented stability of cellulases, for example, [Ch]-based ILs (Ninomiya et al., 2015). Likewise, single-step hydrolysis is preferred as the lignocellulose IL pretreatment is combined with enzymatic hydrolysis to eliminates the stage of cellulose regeneration through washing.

2.2.2 Objective of Experiment

The purpose of this work is to identify the most appropriate cellulase-stabilizing IL to enable lignocellulosic biomass to be hydrolyzed in a single vessel.

2.3 Materials

Tables 2.1, 2.2, and 2.3 list the materials used in this research.

Table 2.1 Consumable items used

No.	Equipment	Aims of usage
1	Pipettes (100 µL, 200 µL, 1 mL)	To add solutions into tubes, microtubes
2	Falcon Tube (15 mL)	To dissolve enzyme and prepare solutions
3	Microcentrifuge tube (2 mL)	To perform enzyme assay and determine total protein
4	Plate/Petri Dish	To contain culture medium (agar)
5	Microplates	Transferring assay solution for absorbance measurement

Table 2.2 Equipment used

No.	Equipment	Usage
1	Weighing balance, Mettler Toledo	To weigh chemicals and reagents
2	Thermomixer, Eppendorf	To incubate and mix the enzyme-IL solution
3	Microplate Spectrophotometer Brand: Multiskan Go™ (Thermo Scientific)	To measure absorbance during determination of enzyme activity and total protein
4	pH meter Brand: Mettler Toledo	To measure pH

Table 2.3 Chemicals and reagents used

No.	Chemicals	Manufacturer
1	1,3-dimethylimidazoliumdimethyl phosphate [DMIM][DMP]	Sigma Aldrich
2	1-ethyl-3-methylimidazolium acetate [EMIM][Ac]	Sigma Aldrich
3	1-ethyl-3-methylimidazolium chloride [EMIM][Cl]	Sigma Aldrich
4	1-ethyl-3-methylimidazolium diethyl phosphate [EMIM][DEP]	Sigma Aldrich
5	3,5-dinitrosalicylic acid	Sigma Aldrich
6	Acetic acid	Friedemann Schmidt
7	Choline hydroxide	Sigma Aldrich
8	Sodium citrate-2-hydrate	Bendosen Laboratory
9	Sodium hydroxide	Bendosen Laboratory
10	Sodium metabisulfite	Fisher Scientific
11	Sodium potassium tartrate	R&M Chemical, UK
12	Sulphuric acid 98%	Fisher scientific
13	Tetrabutyl phosphonium hydroxide	Sigma Aldrich
14	Sodium citrate-2-hydrate	Bendosen Laboratory

2.4 Methodology

2.4.1 Synthesis of Ionic Liquids

[Ch][Ac] was synthesized with slight adjustments using the procedures outlined by Ninomiya et al. (2015). 45.0 wt% of Choline hydroxide [Ch][OH] solution in methanol (Sigma-Aldritch) (100 g) was dispensed dropwise to an equimolar volume of acetic acid (~22.3 g) (Friedemann Schmidt Chemical) in ice-bath. The synthesis was carried out with a round bottom flask with a three-neck that was attached to the condenser and addition funnel. The mixture was left to stir for about 6–12 h before the reaction was stopped. Using the rotary evaporator, methanol was extracted through the vacuum at the time of one hour, at 337 mbar and temperature of 40 °C, while water was evaporated at temperature of 90 °C (2 h, 314 mbar). Using a Freeze dryer (LABCONCO), the resulting residue was vacuum dried to eliminate the residual water. To verify the structure, ^{1}HNMR was used. Choline butanoate [Ch][Bu] was prepared with the same method using butanoic acid in place of acetic acid. Tetrabutyl phosphonium hydroxide was mixed at room temperature with acetic acid to prepare tertabutylphosphonium acetate [TBPH][Ac].

2.4.2 Cellulase Production

Cellulase was prepared at 65% moisture content by fermenting the palm kernel cake (PKC) following the sterilization. The fermentation started with 2% (w/w) of T. reesei spore suspension. Solid-state fermentation (SSF) took place for 7 days at a temperature of 30.0 ± 2 °C. Using citrate buffer (pH 4.8 ± 0.2), the crude enzyme proceeded to extraction followed by the centrifugation. In a multi-step procedure, the enzyme was purified using crossflow filtration. A hollow fiber membrane cartridge was used for ultra-filtration and microfiltration of the cell-free supernatant obtained from centrifugation. For the microfiltration process, a 0.45 μm membrane via 0.011 m² of active surface area was used. Ultra-filtration was performed through ultra-filtration membranes was used (PALL, MWCO 30, and 10 Kd). To determine endo-β-1,4-D-glucanase activity (cellulase) carboxymethyl sodium salt (CMC) was employed as the reactant substance (Salvador et al., 2010).

2.4.3 Stability of Cellulase ILs

The compatibility of ILs with cellulase was investigated. The enzyme was incubated at: 10, 20, 40, 60, 80 and 100% (v/v) of the ILs. As for the control, Tri-Cel was incubated in citrate buffer (50 mM and pH 4.8 ± 0.2). The ILs investigated were 1-ethyl-3-methylimidazolium diethyl phosphate [EMIM][DEP], choline butanoate

[Ch][Bu], choline acetate [Ch][Ac], tetrabutyl phosphonium acetate [TBPH][Ac] and 1,3-dimethyl imidazolium dimethyl phosphate [DMIM][DMP]. CMC hydrolysis was carried out at $45.0 \pm 2.0 °C$ (optimum temperature). For a duration of 6 h, samples were taken every hour, and by using the control (at 100%), which is the enzyme/buffer solution, the activity was described as a residual activity. The activity was evaluated using the dinitrosalicylic acid (DNS) method.

2.4.4 Cellulase Assay

CMC [1.0% (w/v)] was prepared in citrate buffer (pH 4.8 ± 0.2) to assess the activity of endo-β-1,4-D-glucanase. By spectrophotometric quantification of the emitted reducing sugars using DNS, cellulase activity was determined. Substrate solution of 450 μL was prepared with the addition of 35 μL buffer, and the enzyme solution (15 μL) was added. The reaction was terminated after 30 min by adding 1.0 mL of DNS reagent before boiling the solution for 15 min and then cooled before adding water (1.0 mL). The absorbance of the solution was measured at 540 nm. The sugar generated by cellulase was calculated using the glucose standard curve (Fig. 2.1). By measuring different enzyme dilutions, the enzyme concentration that releases approximately 0.5 mg of glucose was recorded. A line was connected for points lower and higher than 0.5 mg, and the enzyme dilution rate (EDR) was defined at 0.5 mg glucose. (Ghose, 1987).

$$CMC = 6.173/EDR \, \text{Unit/mL} \qquad (2.1)$$

In the CMC reaction, the quantity of glucose is generated by 15 μmL in 30 min, where:

0.5 mg glucose = 0.5 mg/(0.18 mg/μmol) × 0.015 mL × 30 min = 6.173 μmol/min/mL.

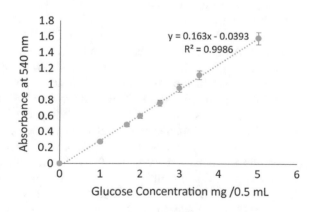

Fig. 2.1 Glucose standard curve for determination of cellulase activity (CMC)

$y = 0.163x - 0.0393$
$R^2 = 0.9986$

Absorbance at 540 nm

Glucose Concentration mg /0.5 mL

Fig. 2.2 Fermentation of palm kernel cake (PKC) by *Trichodermareesei* (7 days)

Under these assay conditions, only 15 mL of the enzyme solution is being used for the response instead of 0.5 mL, so the formula has been modified accordingly.

2.5 Results and Discussion

2.5.1 Cellulase Activity

The enzyme activity was measured using CMC protocol, resulting in 157.872 ± 1.56 CMC units/mL (789.386 ± 7.8 U/gds) following the fermentation process of 7 days (Fig. 2.2). With maximum stability for 24 h at temperatures between 25 and 50 °C, the optimum pH and temperature were 5.00 and 45 °C, respectively.

2.5.2 Stability of Cellulase in ILs

We analyzed the impacts of different types of ILs for six hours on a few concentrations of locally produced cellulase *Tri-Cel*. Six ILs were analyzed for the effect on *Tri-Cel*; [Ch][Ac], [Ch][Bu] [EMIM][Ac], [EMIM][DEP], [DMIM][DMP] and [TBPH][Ac]. Trends in *Tri-Cel* activity can be seen in Fig. 2.3(a–f)

In [Ch][Ac] (Fig. 2.3a), more than six hours with 20% IL/Buffer, locally generated *Tri-Cel* sustained over than 90% of its activity at 10%. The enzyme retained 80 and 85% at 40, 60, and 80% IL/Buffer. At 100% IL/Buffer, after six hours, enzyme activity was detected at 63.15%. In comparison, in 80 and 100% IL/Buffer, [Ch][Bu] (Fig. 2.3b) attained the initial activity at 50%. *Tri-Cel* sustained its activity (>80%) at low concentrations (10 and 20%), whereas in [EMIM][Ac] (Fig. 2.3c) it preserved 85% of the activity at 10–40% IL/Buffer solution. Although 67% activity was recorded at 60% IL/Buffer, high concentrations resulted in a drastic decrease in the activity. In [TBPH][Ac] (Fig. 2.3d), at low concentrations, *Tri-Cel* retained its activity (90%), while less than 20% was identified at higher concentrations of the

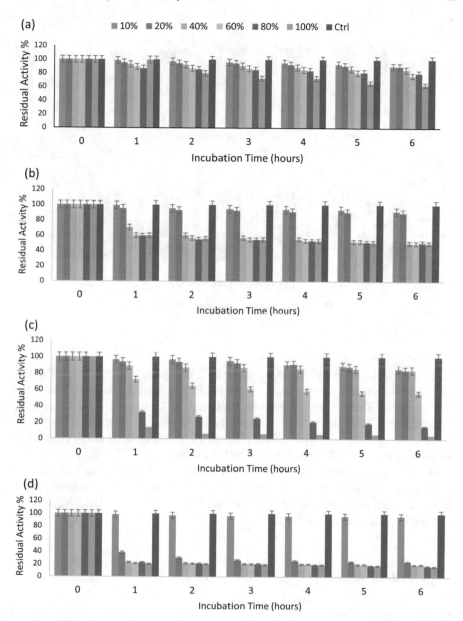

Fig. 2.3 Compatibility of *Tri-Cel* with 6 different ionic liquids (ILs) for a period of 6 h at enzyme optimum conditions (pH 5.0 and 45 °C): **a** [Ch][Ac]. **b** [Ch][Bu]. **c** [EMIM][Ac]. **d** [TBPHA][Ac]. **e** [EMIM][DEP]. **f** [DMIM][DMP]

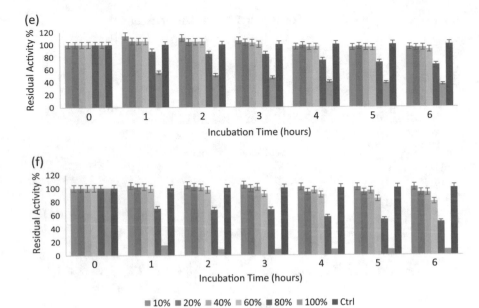

Fig. 2.3 (continued)

IL. Phosphate-based ILs revealed unexpected patterns wherein the [EMIM][DEP] (Fig. 2.3e) stimulated the enzyme in 10–60% IL/Buffer at the first two hours and regulated the activity at 90% in the next six hours. At 80 and 100% IL, the enzyme sustained its activity (70 and 36%), respectively. Similarly, in the initial two hours, a comparable pattern was recorded for [DMIM][DMP] (Fig. 2.3f) at 10–40%, while the activity reduced to 20% in 60% IL/Buffer solution.

In summary, *Tri*-Cel activity was the highest in [Ch][Ac] with an incubation period of six hours, despite being suspended in 100% IL solution. The recorded pattern of low IL concentrations can be in the order: [DMIM][DMP] > [EMIM][DEP] > [Ch][Ac] > [Ch][Bu] > [TBPH][Ac].

2.5.3 Discussion

2.5.3.1 Cellulase Production

Numerous fungal cellulolytic and microbial enzymes have an optimum temperature of 50 °C and optimum activity at pH 4 to pH 6. It was reported (Ni & Tokuda, 2013) that enzyme from *N. koshunensis*; cellobiohydrolase, can function at their best at 45 °C and pH 5.0. Cellulase enzyme from *Trichoderma viride* demonstrated its optimum temperature at 50 °C and at pH 6.0 (Taha et al., 2015). The optimal pH of cellulase agrees with the results of the published studies that the activities of cellulases exhibits their optimal at pH from 4.0 to 7.0 and temperature range of 30 and 40 °C (Pandey et al., 2015). In the acidic range of pH 3.5–6.5 and with the temperature at 40–60 °C, cellulases of the family of *Bacillus* and *Aspergillus* showed their optimal enzyme activity (Assareh et al., 2012; Lin et al., 2012). At pH 5.0 and 45 °C, extracellular cellulase isolated from the marine bacterium *Pseudoalteromonas* sp. had shown the optimal activity. In the crude enzyme blend, the total cellulase (FPase) was 2.11 U/mL and the activity of cellulase (CMCase) was 6.04 U/mL (Trivedi et al., 2013). The latest findings are following the information documented.

2.5.3.2 Cellulase Stability in ILs

It is a fact that ILs digest the cellulose which act as a biocatalysis reaction medium (Swatloski et al., 2002), but residual ILs in the recovered cellulose have been shown to cause enzymatic hydrolysis by inducing activity loss because of the unfolding of the protein (Bose et al., 2010; Turner et al., 2003).

Trivedi et al. (2013) successfully stabilized the extracellular cellulase from marine bacterium *Pseudoalteromonas* sp. in six different type of ILs; 1-ethyl-3-methylimidazolium methanesulfonate [EMIM][CH_3-SO_3], 1-butyl-3-methylimidazolium chloride [BMIM][Cl], 1-butyl-1-methylpyrrolidinium trifluoromethane sulfonate [BMPL][OTF], 1-ethyl-3-methylimidazolium bromide [EMIM][Br], [EMIM][Ac], and 1-butyl-3-methylimidazolium trifluoromethane sulfonate [BMIM][OTF]. When IL solution was used at 5% (v/v), the enzymatic activity was demonstrating the activity higher than 90% for all ILs. In 20% (v/v) IL solution, it was reported that [EMIM][Ac] carries the highest percentage of the enzyme activity which is 94.37% followed by [BMPL][OTF] with the percentage of 80.2%, [BMIM][OTF] (74.69%), [BMIM][Cl] (73.2%), [EMIM][Br] (67%) and [EMIM][CH_3-SO_3] (59%). In addition, the residual activity of the tested enzyme (*Tri-Cel*) in concentrated IL solution (about 60% v/v) of [EMIM][Ac] is comparable with a previous study in which cellulases sustained 86 and 76% of the activity in 5 and 10% of [EMIM][Ac] (Wang et al., 2011b), which validates that cellulases are gradually losing the activity by rising the IL concentration.

ILs with a hydrophobic origin, cosmoropic anion, chaotropic cation and less viscosity in most cases, tend to boost the enzyme's stability and activity. Even so, because of so many conflicting reports, the theory is not generalized (Naushad et al., 2012). In enzymatic hydrolysis system, [DMIM][DMP] and [EMIM][Ac] were both investigated and revealed that when IL concentration higher than 40% resulted in the cellulase deactivation, endoglucanase sustained its activity (50%) in a solution of 90% (v/v) [DMIM][DMP] (Wahlström et al., 2012). Similarly, after one hour, cellulase sustained about 40% of the activity in [EMIM][Ac] (Ebner et al., 2014). Fukaya et al., (2008) suggested that in enzymatic catalysis, the anionic element of ILs portrayed an important role, whereas a single-step continuous process is used for biomass treatment and saccharification, cellulase in ILs is regarded as a viable alternative. *Tri-Cel* could therefore function as an excellent biocatalysis for biomass hydrolysis since it is generated locally at minimal cost by optimizing the waste utilization from agro-industrial.

2.6 Conclusion

Tri-Cel has good activity and stability. Of all ILs that were evaluated in this research, [Ch][Bu] and also [Ch][Ac] provided excellent media for the *Tri-Cel*-ILs system. This method is promising on the basis of the analysis and recommended for a one-step process for lignocellulose treatment and hydrolysis. ILs with cholinium cations have shown good compatibility with cellulase enzyme and could be utilized in future studies.

Acknowledgements The author would like to acknowledge the Ministry of Education Malaysia for awarding the FRGS Grant (Grant no. FRGS-13-088-0329) and the Research Management Centre, IIUM for the grant [RMCG20-021-0021].

References

Asakawa, A., Kohara, M., Sasaki, C., Asada, C., & Nakamura, Y. (2015). Comparison of choline acetate ionic liquid pretreatment with various pretreatments for enhancing the enzymatic saccharification of sugarcane bagasse. *Industrial Crops and Products, 71,* 147–152. http://dx.doi.org/10.1016/j.indcrop.2015.03.073.

Assareh, R., Zahiri, H. S., Noghabi, K. A., & Aminzadeh, S. (2012). Characterization of the newly isolated Geobacillus sp. T1, the efficient cellulase-producer on untreated barley and wheat straws. *Bioresource Technology, 120,* 99–105.

Bose, S., Armstrong, D. W., & Petrich, J. W. (2010). Enzyme-catalyzed hydrolysis of cellulose in ionic liquids: A green approach toward the production of biofuels. *Journal of Physical Chemistry B, 114*(24), 8221–8227. Retrieved from http://www.scopus.com/inward/record.url?eid=2-s2.0-77953744655&partnerID=40&md5=472f4e59d92c06002b34bacf86306e04.

De Souza Mesquita, L. M., Murador, D. C., & De Rosso, V. V. (2019). Application of ionic liquid solvents in the food industry. In *Encyclopedia of ionic liquids* (pp. 1–16). Springer Singapore. https://doi.org/10.1007/978-981-10-6739-6_8-1.

Ebner, G., Vejdovszky, P., Wahlström, R., Suurnäkki, A., Schrems, M., Kosma, P., … Potthast, A. (2014). The effect of 1-ethyl-3-methylimidazolium acetate on the enzymatic degradation of cellulose. *Journal of Molecular Catalysis B: Enzymatic, 99,* 121–129. https://doi.org/10.1016/j.molcatb.2013.11.001.

Elgharbawy, A. A., Alam, M. Z., Moniruzzaman, M., & Goto, M. (2016). Ionic liquid pretreatment as emerging approaches for enhanced enzymatic hydrolysis of lignocellulosic biomass. *Biochemical Engineering Journal, 109,* 252–267. https://doi.org/10.1016/j.bej.2016.01.021.

Fu, D., Mazza, G., & Tamaki, Y. (2010). Lignin extraction from straw by ionic liquids and enzymatic hydrolysis of the cellulosic residues. *Journal of Agricultural and Food Chemistry, 58*(5), 2915–2922.

Fukaya, Y., Hayashi, K., Wada, M., & Ohno, H. (2008). Cellulose dissolution with polar ionic liquids under mild conditions: Required factors for anions. *Green Chemistry, 10*(1), 44–46.

Ghose, T. K. (1987). Measurement of cellulase activities. *Pure and Applied Chemistry, 59*(2), 257–268. https://doi.org/10.1351/pac198759020257.

Gunny, A. A. N., Arbain, D., Edwin Gumba, R., Jong, B. C., & Jamal, P. (2014). Potential halophilic cellulases for in situ enzymatic saccharification of ionic liquids pretreated ligno-celluloses. *Bioresource Technology, 155*(0), 177–181. http://dx.doi.org/10.1016/j.biortech.2013.12.101.

Ibrahim, F., Moniruzzaman, M., Yusup, S., & Uemura, Y. (2015). Dissolution of cellulose with ionic liquid in pressurized cell. *Journal of Molecular Liquids, 211,* 370–372. https://doi.org/10.1016/j.molliq.2015.07.041.

Lin, L., Kan, X., Yan, H., & Wang, D. (2012). Characterization of extracellular cellulose-degrading enzymes from Bacillus thuringiensis strains. *Electronic Journal of Biotechnology, 15*(3), 1–7. https://doi.org/10.2225/vol15-issue3-fulltext-1.

Moniruzzaman, M., Mahmood, H., Kamiya, N., Yusup, S., & Goto, M. (2015). Activity and stability of enzyme immobilized with ionic liquid based polymer materials. *Journal of Engineering Science and Technology,* 60–69.

Naushad, M., ALOthman, Z. A., Khan, A. B., & Ali, M. (2012). Effect of ionic liquid on activity, stability, and structure of enzymes: A review. *International Journal of Biological Macromolecules, 51*(4), 555–560. https://doi.org/10.1016/j.ijbiomac.2012.06.020.

Ni, J., & Tokuda, G. (2013). Lignocellulose-degrading enzymes from termites and their symbiotic microbiota. *Biotechnology Advances, 31*(6), 838–850. http://dx.doi.org/10.1016/j.biotechadv.2013.04.005.

Ninomiya, K., Inoue, K., Aomori, Y., Ohnishi, A., Ogino, C., Shimizu, N., & Takahashi, K. (2015). Characterization of fractionated biomass component and recovered ionic liquid during repeated process of cholinium ionic liquid-assisted pretreatment and fractionation. *Chemical Engineering Journal, 259*(0), 323–329. http://dx.doi.org/10.1016/j.cej.2014.07.122.

Pandey, S., Srivastava, M., Shahid, M., Kumar, V., Singh, A., Trivedi, S., & Srivastava, Y. K. (2015). Trichoderma species cellulases produced by solid state fermentation. *Journal of Data Mining in Genomics & Proteomics, 2015.*

Salvador, Â. C., Santos, M. D. C., & Saraiva, Ja. (2010). Effect of the ionic liquid [bmim]Cl and high pressure on the activity of cellulase. *Green Chemistry, 12*(4), 632. https://doi.org/10.1039/b918879g.

Shi, J., Gladden, J. M., Sathitsuksanoh, N., Kambam, P., Sandoval, L., Mitra, D., ... Singh, S. (2013). One-pot ionic liquid pretreatment and saccharification of switchgrass. *Green Chemistry, 15*(9), 2579. https://doi.org/10.1039/c3gc40545a.

Swatloski, R. P., Spear, S. K., Holbrey, J. D., & Rogers, R. D. (2002). Dissolution of cellose with ionic liquids. *Journal of the American Chemical Society, 124*(18), 4974–4975.

Taha, A. S. J., Taha, A. J., & Faisal, Z. G. (2015). Purification and kinetic study on cellulase produced by local Trichoderma viride. *Nature and Science, 13*(1), 87–90.

Tan, H. T., Lee, K. T., & Mohamed, A. R. (2011). Pretreatment of lignocellulosic palm biomass using a solvent-ionic liquid [BMIM]Cl for glucose recovery: An optimisation study using response surface methodology. *Carbohydrate Polymers, 83*(4), 1862–1868. http://dx.doi.org/10.1016/j.carbpol.2010.10.052.

Trivedi, N., Gupta, V., Reddy, C. R. K., & Jha, B. (2013). Detection of ionic liquid stable cellulase produced by the marine bacterium pseudoalteromonas sp. isolated from brown alga Sargassum polycystum C. Agardh. *Bioresource Technology, 132,* 313–319. https://doi.org/10.1016/j.biortech.2013.01.040.

Turner, M. B., Spear, S. K., Huddleston, J. G., Holbrey, J. D., & Rogers, R. D. (2003). Ionic liquid salt-induced inactivation and unfolding of cellulase from Trichoderma reesei. *Green Chemistry, 5*(4), 443–447.

Wahlström, R., Rovio, S., & Suurnäkki, A. (2012). Partial enzymatic hydrolysis of microcrystalline cellulose in ionic liquids by Trichoderma reesei endoglucanases. *RSC Advances, 2*(10), 4472–4480. http://dx.doi.org/10.1039/C2RA01299E.

Wang, Y., Radosevich, M., Hayes, D., & Labbé, N. (2011a). Compatible ionic liquid-cellulases system for hydrolysis of lignocellulosic biomass. *Biotechnology and Bioengineering, 108*(5), 1042–1048. Retrieved from http://www.scopus.com/inward/record.url?eid=2-s2.0-799 53144636&partnerID=40&md5=d7db5ac4ddc5b768c7757d5259f1b978.

Wang, P., Yu, H., Zhan, S., & Wang, S. (2011b). Catalytic hydrolysis of lignocellulosic biomass into 5-hydroxymethylfurfural in ionic liquid. *Bioresource Technology, 102*(5), 4179–4183. http://dx.doi.org/10.1016/j.biortech.2010.12.073.

Zhao, H., Jones, C. L., Baker, G. A., Xia, S., Olubajo, O., & Person, V. N. (2009). Regenerating cellulose from ionic liquids for an accelerated enzymatic hydrolysis. *Journal of Biotechnology, 139*(1), 47–54. http://dx.doi.org/10.1016/j.jbiotec.2008.08.009.

Chapter 3
Role of Ionic Liquids in the Processing of Lignocellulosic Biomass

**Amal A. M. Elgharbawy⊚, Sharifah Shahira Syed Putra,
Md Zahangir Alam, Muhammad Moniruzzaman,
Nassereldeen Ahmad Kabbashi, and Parveen Jamal**

Abstract This chapter discusses the treatment of palm oil empty fruit bunch using ionic liquids (ILs) pretreatment. By mixing IL and cellulase enzyme (IL-E) in a single pot, the empty fruit bunch (EFB) was pretreated with simultaneous fermentation. Choline acetate [Ch][Ac], which has excellent biological compatibility and renewability, has been used for pretreatment. Chemical analysis, electron scanning microscopy (SEM), and Fourier transform infrared spectroscopy (FTIR) were used to characterize the EFB and its hydrolysate. After 24 and 48 h of enzymatic hydrolysis, sugar yield improved from 0.058 g/g EFB in the crude (untreated) sample to 0.283 and 0.62.06 g/g in the IL-E phase. EFB hydrolysate demonstrates suitability for the production of ethanol (EtOH) with a yield of 0.275 g EtOH/g EFB in the presence of [Ch][Ac] without additional nutrients, compared to the low yield without IL pretreatment.

Keywords Ionic liquid · Lignocellulose · Pretreatment · Cellulase · Hydrolysis

3.1 Introduction

The palm oil industry produces noteworthy amounts of lignocellulosic palm oil biomass (OPB). Approximately 21.625 tonnes of biomass per year can be generated per hectare of oil palm plantation. Chew and Bhatia (2008) stated that 50.31%

A. A. M. Elgharbawy (✉)
International Institute for Halal Research and Training (INHART), International Islamic University Malaysia (IIUM), Gombak, Malaysia
e-mail: amalgh@iium.edu.my

S. S. S. Putra
Chemistry Department, Universiti Malaya, Kuala Lumpur, Malaysia

A. A. M. Elgharbawy · M. Z. Alam · N. A. Kabbashi · P. Jamal
Department of Biotechnology Engineering, International Islamic University Malaysia (IIUM), Kuala lumpur P.O. Box 10, 50728, Malaysia

M. Moniruzzaman
Chemical Engineering Department, Universiti Teknologi PETRONAS, Seri Iskandar, Malaysia

(fronds) and 20.44% (EFB) is reported to have the highest levels of cellulose. The key components of oil palm EFB (OPEFB) are lignin (17.6%) and hollow cellulose (82.45%). OPEFB is high in carbohydrate content which can be derived for producing mannose, ethanol, animal feed, and xylose as it is known to be a cheap source of carbohydrates (Ishola et al., 2014). Thanks to their composition, these agro-industrial substances are alternative tools for the development of biofuels and can be used for the production of sugar and the bioconversion of different forms of products (Zainan et al., 2013). While cellulose is the most important component of lignocellulosic substances, it is covered by hemicellulose and lignin (Laureano-Perez et al., 2005), which complicates the process of hydrolysis using cellulases. As a result, the pretreatment of lignocellulose for hydrolysis is the most critical stage in the effective use of biomass (Hahn-Hägerdal et al., 2006).

ILs have been implemented as useful solvents for the pretreatment of lignocellulose, enabling hydrogen attachments to be disrupted and allowing molecules' exposure to cellulase hydrolysis. Despite this, ILs can induce a loss of cellulase function, which is controlled by the analysis to ensure that cellulase can tolerate the effects of ILs. As reported by Wang and colleagues (Wang et al., 2011), who inspected ILs with yellow polar biomass saccharification at 15% [EMIM]Ac, some commercial cellulases were active in a few ILs. Switchgrass treatment was also conveyed in 10–20% of [EMIM]Ac before enzymatic hydrolysis (Shi et al., 2013). Moreover, in order to detect IL-enzyme consistency in cholinium-based ILs, several researchers have researched cellulase stability (Ninomiya et al., 2015a). On the other hand, single-pot hydrolysis is an interesting bioethanol conversion process that combines lignocellulosic substance and IL treatment. Simple in situ lignocellulose conversion could result from recovery, which could help develop an integrated ethanol production system.

Researchers have used a number of methods to extract EFB sugar, including enzymatic treatment, sodium hydroxide and steam (Choi et al., 2013), high temperature diluted sulphuric acid (Chong et al., 2013) and ILs treatment (Ninomiya et al., 2015a). Treatment with acid and alkali, on the other hand, induces the release of HMF and furfural, which are both toxic and hinder microbial growth. Many ILs are versatile substances that create less by-products while preserving enzyme stability and allowing microorganisms to grow naturally (Reddy, 2015).

Enzymatic hydrolysis precedes the extraction of sugar from lignocellulose by a multiple-step process that includes cellulose regeneration by washing, according to the majority of previous research. For instance, *Nothofagus pumilio* (Lienqueo et al., 2016), rice straw (Poornejad et al., 2014), bagasse (Ninomiya et al., 2015b), and *Eucalyptus globulus* Labill are fermented to generate ethanol from the hydrolysate. This research, however, focused on the synthesis of IL and cellulase in a single vessel by using EFB. In addition, a detailed analysis was conducted before and after hydrolysis to assess the hydrolysate's suitability for yeast fermentation. Predictably, this strategy will provide a clear understanding of the immediate step effect of IL-cellulase (IL-E) on the conversion of lignocellulose into sugar and thus into ethanol.

3.2 A Brief Perspective of Ionic Liquid

Scientists working in many different fields of study have come to recognize the peculiar characteristics of ILs over the past decade and to acknowledge the promise of these new materials. They are non-toxic and reusable replacement to volatile organic compounds (VOCs), which attract researchers and professional interests as potential pretreatment solvents for lignocellulosic biomass prior to further processing (Moniruzzaman & Goto, 2018). The ability of ammonium salts to break down cellulose was first acknowledged in 1934. The use of ILs for biomass production (which typically involves cellulose) is now a well-established and well-researched field (Graenacher, 1934; Macfarlane et al., 2017).

Using ILs as solvents or co-solvents can solve substrates' low solubility and materials during wood delignification in aqueous solutions. This is primarily due to enzyme entry issues within concrete substrates and low solubility of substrates and materials, such as lignin, which dignifies slowly in aqueous systems (Sousa et al., 2009). Unfortunately, some ILs, such as hydrophilic ILs can have adverse effects on enzyme structure which can cause the deactivation of the enzyme when used for enzyme delignification. By increasing the solubility of substrates and materials, the overall process efficacy can be increased. Therefore, such an effect can be balanced. Prior to enzymatic hydrolysis and delignification, it is widely acknowledged that ILs can be used to successfully pretreat lignocellulosic materials (Elgharbawy et al., 2016a).

3.2.1 Pretreatment Process for Lignocellulosic Biomass

Pretreatment, enzyme hydrolysis, fermentation and ethanol separation are the four steps involved in converting lignocellulose into bioethanol. To effectively extract hemicellulose and lignin, inexpensive approach that improves cellulose hydrolysis is used as the first step in the pretreatment process. Pretreatment is the process of modifying biomass in order to speed up and increase the yields of cellulose and hemicellulose enzymatic hydrolysis. This involves removing lignin, which increases surface area while decreasing cellulose crystallinity. In contrast to pure lignin, extracting natural lignin in wood is more difficult owing to its complex structure and strong intramolecular reactions with lignocellulose. Due to the presence of strong covalent bonds, it is more vulnerable to breakdown than other lignocellulosic elements. ILs are effective in treating hemicellulose and lignin, as the crystallinity of cellulose is decreased (Elgharbawy et al., 2016b). Sun et al. (2009) observed that a wide variety of substrates, like softwood and hardwood, can be dissolved using ILs.

Enzyme adsorption into cellulose particles and enzyme-substratum complexes (ES) aggregation that are further correlated with several enzymes and substrate factors are basically two stages of cellulose enzyme-catalyzed hydrolysis (Chandra et al., 2012). Compound inhibition (cellobiose and D-glucose), adsorption, synergy and thermal stability are factors to consider when it comes to enzymes. Meanwhile, external and internal cellulose surface proximity, hemicellulose and lignin, degree of polymerisation (DP), and cellulose crystallinity are substrate related variables (Zhao et al., 2009). Zhao et al. (2009) hypothesized that cellulose is more accessible to regenerated cellulose since IL-regenerated cellulose is less crystalline compare to untreated cellulose and prevents quicker hydrolysis of the enzyme. Elgharbawy et al. (2016c) stated that during the fermentation process, lignocellulose biomass, cellulose, and hemicellulose are predominantly hydrolyzed to sugar monomers and eventually converted to alcohol, hydrogen, or methane.

3.2.2 Ionic Liquid–Cellulase Compatible Systems

Multi-stage pretreatment of lignocellulosic materials with different ILs, accompanied by IL removal and recovery of cellulose, followed by hydrolysis process using enzymes was highlighted in several reports (Bian et al., 2014; Elgharbawy et al., 2016b). Researchers have also studied hydrolysis in single-stage processing of bioethanol production that combines lignocellulosic substances pretreatment and enzymatic hydrolysis.

This is because it was possible to customize ILs to adapt to a specific reaction. In addition, the thermal tolerance of the enzymes in ILs enables a reaction to being carried out at higher temperatures. Researchers tend to regenerate IL-biomass cellulose before enzymatic hydrolysis to mitigate the adverse effect of ILs on enzymatic hydrolysis (Elgharbawy et al., 2016c; Moniruzzaman et al., 2010).

Elgharbawy and colleagues (Elgharbawy et al., 2016a) suggest that combining ILs with cellulase in a system necessarily requires proper component selection as well as reaction control conditions. The following are typical characteristics of the enzyme with ILs (IL-E):

i.   Using a broad molecular structure to reduce the basic nature and nucleophilicity of H-bonding.
ii. It has a wide range of ether and/or hydroxyl groups to improve water affinity, H-bond basicity for mild enzyme behaviors and enzyme viscosity (Moniruzzaman et al., 2010).

3.3   Objective of Experiment

The primary aim is to design a cellulase-compatible IL system for the development of bioethanol from EFB in a single reactor. Therefore, the primary aim of the suggested study is as follows:

1.   To select a cellulase compatible with ionic liquid (IL-E), which can be used for single-step EFB hydrolysis.

2. To determine the effect of IL-E on the characteristics of the substrate on untreated and treated EFB.
3. To maximize factors that affect the E-IL mechanism and biomass dissolution in a single stage.
4. To enhance bioethanol fermentation by mixing fermentable sugars derived from IL-E with yeast (*Saccharomyces cerevisiae*).
5. To investigate the kinetics of EFB saccharification and bioethanol production in the IL-E system.

3.4 Materials

Tables 3.1, 3.2 and 3.3 show all materials used in this study.

Table 3.1 Consumable items used

No.	Equipment	Aims of usage
1	Pipettes (100 µL, 200 µL, 1 mL)	To add solutions into tubes, microtubes
2	Felcon tube (15 mL)	To dissolve enzyme and prepare solutions
3	Microcentrifuge tube (2 ml)	To perform enzyme assay
4	Whatmann No. 1 filter paper	To monitor the activity in filter paper units (FPU)
5	Microplates	Transferring assay solution for absorbance
6	Erlenmeyer flask (250 mL)	To autoclave 100 mL of the hydrolysate

Table 3.2 Equipment used

No.	Equipment	Usage
1	Weighing balance, Mettler Toledo	To weigh chemicals and reagents
2	Thermomixer, Eppendorf	To incubate and mix the enzyme-IL solution
3	Microplate Spectrophotometer Brand: Multiskan Go^{TM} (Thermo Scientific)	To measure absorbance during determination of enzyme activity and total protein
4	pH meter Brand: Mettler Toledo	To measure pH
5	High-performance liquid chromatography (HPLC)	To determine monosaccharides and oligosaccharides
6	Fourier Transform Infrared Spectroscopy (FTIR)	To determine the effect of pretreatment on the structure of the EFB
7	Absorption Spectrometry (AAS)	To determined iron (Fe), zinc (Zn) and manganese (Mn)
8	Screening electron microscopy (SEM)	To characterize native (untreated) and treated samples

Table 3.3 Chemicals and reagents used

No.	Chemicals	Manufacturer
1	Sodium Citrate-2-hydrate	Bendosen
2	Carboxymethyl Cellulose Sodium Salt (CMC)	Sigma Aldrich
3	Sulphuric Acid (98%)	Fisher Scientific
4	Phenol	Sigma Aldrich
5	Acetic Acid	Friedemann
6	Choline hydroxide	Sigma Aldrich
7	Sodium butyrate	Sigma Aldrich
8	Sodium hydroxide	Bendosen
9	Sodium Potassium Tartrate	R&M Chemical

3.5   Methodology

3.5.1   Production of PKC-Cel

As mentioned earlier in our study, locally produced cellulase was obtained as a PKC fermentation product (Elgharbawy et al., 2016a). For the primary medium used for processing, the cellulase enzyme was labeled 'PKC-Cel'. The base medium was a solid-state fermentation of palm kernel cake. Fermentation occurred after inoculating samples of *Trichoderma reesei* (RUTC30) suspension for seven days at room temperature (30.0 ± 2 °C). The crude enzyme was extracted with a sodium citrate buffer (pH 4.8 ± 0.2) before centrifuge, micro-filtration, and ultra-filtration. The activity of the retentate-containing enzyme was determined. Whatman No. 1 was used as a substrate material to monitor the hydrolysis of filter paper units (FPU). One unit of FPU was estimated to release 1 mol per mL enzyme per minute of glucose. A glucose standard curve was used to calculate the amount of glucose produced. Sodium carboxymethyl cellulose (CMC) has been studied as a substrate for endo-b-1,4-D-glucanase (Salvador et al., 2010). As Ghose (1987) states, one unit of CMCase is defined as the quantity of enzyme produced by 1 mol of glucose per minute. On both filter paper and CMCase, the assay was performed.

3.5.2   Enzymatic Hydrolysis of EFB

The ground EFB was seived into three different particle dimensions, as suggested by the arrangement (200, 450, and 600 μm) (Retsch AS 300). For EFB pretreatment, separate biomass loading percentages of volatile solids (percent VS) were used by ignition the sample at 550 °C for 30 min. To facilitate the pretreatment, 1.875 g of EFB were weighted in a glass vial, and 5 g of IL was added, followed by a 60 min incubation at 75 °C. In order to achieve a 10% IL buffer solution (v/v), the sample was mixed with pH 4.8 of sodium citrate buffer after incubation before cooling to room temperature. After that, 45 FPU/g EFB of ultra-filtered (UF) cellulase was prepared. The DNS assay was used to measure the released sugar concentration after 48 h of enzyme reaction. Elgharbawy et al. (2016a) and (2016b) have described methods and techniques in prior studies.

3.5.3 Characterization of the Native and Treated EFB for IL-E Evaluation

The EFB's chemical constituents were identified with some modifications in accordance with the criteria mentioned in the NREL test techniques, as published by Ninomiya et al. (2015a).

3.5.4 Structural Carbohydrates and Sugar Determination

By mixing 2 mL of 72% (v/v) H_2SO_4 aqueous solution with 0.1 g of the sample for 2 h at room temperature, then mixed with 75 mL of water and autoclaving around 121 °C for 15 min, hemicellulose and lignin contents were determined. The solid residue was used to calculate the weight at 100 °C after 12 h of drying to determine acid-insoluble lignin (AIL) after filtering the acid-diluted hydrolysate. The amount of acid-soluble lignin (ASL) was calculated at 205 nm with a UV absorbance of 110 L/g/cm.

The total of AIL and ASL was calculated as the sum of all lignin (Ninomiya et al., 2015c). As seen in Eq (3.1), the quantity of hemicellulose was calculated in terms of the content of xylose:

$$
Hemicellulose\left(\frac{g}{g}\,biomass\right) =
$$

$$
Xylose\,concentration\left(\frac{g}{L}\right) \times \frac{volume\,of\,sample}{weight\,of\,dry\,EFB} \times Correction\,factor\left(\frac{132}{150}\right) \qquad (3.1)
$$

By using monomeric sugar concentration such as anhydrous xylose and arabinose correction factors (0.88 or 132/150), the ratios obtained from the respective polymeric sugars can be determined (Nieves et al., 2016). The quality of untreated and refined EFB cellulose was measured using standard microcrystalline cellulose using the anthrone technique (Updegraff, 1969). Using phenol-sulfuric acid analysis, the quantity of biomass carbohydrates was calculated (Dubois et al., 1956).

The total soluble sugar in the solution was evaluated by dissolving 100 mg of EFB in 100 mL of distilled water, and the total free sugar was collected and filtered using filter paper (Whatman No. 1).

The acquired filtrate was analyzed using the phenol-sulfuric technique. Following the report of Salvador et al. (2010), total glucose, galactose, mannose, and arabinose reduction in sugar (TRS) were determined. A COSMOSIL SUGAR-D column (4.6 mm l. D. × 250 mm) for high-performance liquid chromatography (HPLC) comprised of high purity porous spherical silica gel was used to analyze monosaccharides and oligosaccharides. Primary and tertiary amines are the main components of the stationary phase. In hydrolyzed samples, arabinose, glucose, mannose, galactose, and xylose occur on the sugar solution. Syringe filter nylon (0.2 μm) was used to purify each sample (1.0 mL) into a transparent HPLC glass vial. Deionized water was used as a solvent to make four different standards of each sugar. A syringe filter was used to transfer standard dilutions to HPLC vials. In the Waters 600 (Hampton, USA) system, HPLC was used with a refractive index detector (RID) fitted with an acetonitrile mobile phase with deionized water ratio of 70:30, 75:25, and 80:20 in a set of 1.0 mL/min flow rate at 30 °C. The mobile phase was passed through a membrane filter before running the HPLC to prevent column blockage.

The concentration of acetic acid (CH_3COOH) was calculated using an Agilent 1200 HPLC process with RID and column of REZEX ROA (Phenomenex, USA) (California, USA). With isocratic elution, sulfuric acid (5 mmol/L) was used as a mobile step at a 0.6 mL/min flow rate at ambient temperature. Using variety of acetic acid levels with 20 μL of sample injection, standard curves were produced. To maximize the main column's longevity, an identical packed guard column was used. Institute of Systems Biology (INBIOSIS) of Universiti Kebangsaan Malaysia (UKM), situated in Selangor (Bangi), Malaysia, kindly gathered the results of the research.

3.5.4.1 Morphological Structural Changes

The samples of treated and native (untreated) were characterized using SEM. Both samples were stored at −80 °C after being dried using a freeze dryer (LABCONCO). Dry samples were used for observation using the Hitachi SU1510 SEM Back-Scatter Detector (BSE) Mode at 500X, 1000X and 5000X magnifications for both treated and untreated samples. The analysis was carried out at UKM-MTDC Smart Technology Center (Quasi-S Sdn. Bhd), Bangi, Malaysia.

3.5.4.2 Fourier Transform Infrared Spectroscopy (FTIR)

The hydrolysate, IL-treated, and raw EFB were measured using FTIR after 72 h frozen dry in the range of 4000–5000 cm^{-1} at 4 cm^{-1} resolution with 16 scans. In the absorbance mode, the spectral results were reported as a function of the wavenumber.

3.5.4.3 Swelling Capacity

For one hour, a dry sample (0.1 g) was put in a non-woven fiber bag and immersed in distilled water to calculate pretreated and native EFB swelling capacity (Noori & Karimi, 2016). The swelling capacity was estimated by calculating the different dry weight (W_1) and swollen weight (W_2) materials using Eq. (3.2):

$$Swelling\ Capacity = (W_2 - W_1)/W_1 \qquad (3.2)$$

3.5.4.4 Enzyme Adsorption and Desorption

In 15 mL citrate buffered centrifuge tubes, 400 FPU/g cellulase was diluted up to 5.0 mL with 0.05 g of substrate samples (native or pretreated) to 50 mM in pH 4.8. For two hours, all tubes were mixed at 100 rpm and incubated around 4 °C. After centrifugtion, the Bradford assay was used to measure the supernatant protein content of unabsorbed cellulase (15 min, 4000 rpm, 4 °C). Prepared samples were diluted at 4 °C with a 5 mL citrate buffer for enzyme desorption estimation and incubated for 2 h at 4 °C for determination of protein content after centrifugation. Equation 3.3 show the calculation of total desorption of cellulase percentage:

$$Cellulase\ Desorption\ (\%) = (C - D)/Dx100 \qquad (3.3)$$

where after 2 h, C is the total amount of adsorbed cellulase to solid pretreated (mg/g), and D is the sum of unadsorbed proteins (mg/g) (Noori & Karimi, 2016).

3.5.4.5 Crystallinity Index

The effect of EFB structure pretreatment was determined using FTIR adsorption analysis. The 1430 cm^{-1} absorption band represents hydrolysis-resistant cellulose I, while the 896 cm^{-1} band is easily hydrolyzed by cellulose II (Kljun et al., 2011).

3.5.4.6 Determination of Total Nitrogen (TN) and Total Phosphorus (TP)

After hydrolysis, the overall nitrogen content of the samples was determined using HACH technique 10071 (Test 'N Tube Vials). The molybdovanadate reagent was employed to calculate phosphorus in the samples.

3.5.4.7 Determination of Minerals

Referred to the UNIPEQ-MTDC Technology Centre, UKM (Ref: ULUKM/1787/16), in-house Method (No. STP/Chem/A13 Microwave Digestion-AAS) was used to evaluate the content of calcium (Ca), sodium (Na), copper (Cu) and potassium (K). By using atomic absorption spectrophotometry (AAS), zinc (Zn), iron (Fe) and manganese (Mn) were investigated. Including the standards, 0.2% nitric acid was used to prepare all samples.

3.5.4.8 Determination of Total Phenolic Compounds (TPC)

The MultiskanTM GO Microplate Spectrophotometer is used to determine total phenolic compounds (TPC) in specimens at 760 nm by using the Folin-Ciocalteu method (Wolfe et al., 2003). In mg per 100 g of sample, the standard curve of gallic acid solutions prepared was used to measure the absorption of gallic acid. For each sample, triplicate assay samples were analyzed.

3.5.4.9 Determination of Furfural

To assess the furfural content, the phenylhydrazine solution technique was used and the absorbance was recorded at 446 nm (Zarei, 2009). The exact methods were followed for the standard preparation, except the sample solution was substituted with various concentrations of furfural solution (0.5, 1.5, and 2.0 g/L).

3.5.4.10 Fermentation to Ethanol

On the autoclaved hydrolysate (100 mL), fermentation was performed in a 250 mL Erlenmeyer flask. After chilling, 2.5 mL of prepared inoculum (*Saccharomyces cerevisiae*) was inoculated, tightly covered with a cotton lid, and incubated at 30 °C at 150 rpm. After 72 h, the broth was centrifuged at 8000 rpm for 15 min, and the EtOH concentration was evaluated using chromic acid procedure (Caputi et al., 1968) and then subjected to GC/MS. In the GC/MS, a DB-WAX column (122–7032) was used, as well as an Agilent CTC-PAL autosampler (7890A GC and 5975C MS) fitted with a headspace sampling module capillary column (Agilent Technologies, Wokingham, UK) using 25 cm/s helium gas carrier at 150 °C. Using the DNS assay, the average sugar consumption was monitored for 84 h. To extract ILs from EtOH, distillation was used, where EtOH is obtained at the receiver vessel.

3.6 Results and Discussion

3.6.1 Characterization of the Native (Untreated) and Treated Substrate to Evaluate the Effect of IL-E

3.6.1.1 Pretreatment and Saccharification of EFB

Table 3.4 displays the effects of a 24-hour IL hydrolysis to determine which solvent can achieve the maximum level of sugar. To be subjected to EFB hydrolysis, [Ch][Ac] and [Ch][Bu] are used in sample pretreatment. Conversion variations between pretreated and native EFB demonstrated that the structure has been weakened by IL pretreatment and has increased the enzyme's sensitivity towards the substrate.

During pretreatment, [Ch][Ac] and [Ch][Bu] were found to be completely aligned with PKC-Cel in this study, as confirmed by a decrease in lignin level. Likewise, with the presence of ILs in the environment, there was a rise in saccharification. An amount of 0.14 ± 0.05 g of glucose was obtained with [Ch][Ac] from 1 g of EFB. When [Ch][Bu] was used, the production was slightly lower.

Table 3.4 EFB-pretreated and hydrolyzed sample analysis in two IL-compatible system

Entry	Acid-soluble lignin % (w/w)	Acid-insoluble lignin % (w/w)	Total lignin % (w/w)	Cellulose content % (w/w)
Content after pretreatment with IL compared with native EFB[a]				
Native EFB	1.14 ± 0.1	22.5 ± 0.5	23.6 ± 0.6	2.1 ± 1.01
[Ch][Ac]	0.91 ± 0.05	15.9 ± 0.3	16.8 ± 0.35	32.6 ± 1.5
[Ch][Bu]	1.10 ± 0.05	17.0 ± 0.5	18.1 ± 0.55	26.2 ± 0.54
Entry	Converted hemicelluloses % (w/w)	Residual hemicelluloses content % (w/w)	Total reducing sugar % (g/L)	Glucose (g/L)
Content after hydrolysis in the presence of IL[b]				
Native EFB	0.6 ± 0.00	23.1 ± 0.87	10.0 ± 0.1	0.12 ± 0.1
[Ch][Ac]	21.4 ± 0.2	10.8 ± 1.67	31.7 ± 0.1	16.6 ± 0.5
[Ch][Bu]	15.7 ± 0.2	12.3 ± 2.05	30.1 ± 0.1	15.1 ± 0.1

[a]Conditions: 60 min pretreatment, 75 °C
[b]Conditions: 10% IL, 50 °C, 45 FPU/g EFB, 24 h enzymatic hydrolysis

3.6.1.2 Changes in Cellulose, Hemicellulose, and Lignin

The hydrolysis-accessible native EFB cellulose content was estimated around 2.1 ± 1.0% (w/w) and when ILs were applied, the cellulose content increased. Table 3.4 reveals that [Ch][Ac] has the highest cellulose, with 32.6 ± 1.5% (w/w) (accessible for hydrolysis after 24 h). Without pretreatment with ILs, improvements in hemicellulose tended to be null.

The overall native EFB lignin content was estimated at 23.6 ± 0.6% (w/w) with a further substantial decrease of 16.81 ± 0.35% (w/w) in [Ch][Ac]. [Ch][Ac] was 31.7 ± 0.1 g/L, while the total native EFB hydrolyzed sugar was 10.02 ± 0.1 g/L. For [Ch][Ac] after hydrolysis, the maximum detectable glucose concentration was 16.57 0.5 g/L in specific case.

3.6.1.3 Scanning Electron Microscopy (SEM) for Morphological Structural Changes

SEM was used to investigate the structure and morphology. Figures 3.1 and 3.2 represent EFB native samples and those [Ch][Ac] and [Ch][Bu] treated micrograph, respectively. The study of the pretreated samples showed slight variations, while sample structure in both IL yielded identical findings. At first glance, a distorted surface was noticeable for both ILs, which seemed to be robust and unbending in contrast to the native EFB. Cellulose strands were detected after an hour by examining the IL-pretreated samples. Figure 3.1 displays comparable EFB images of [Ch][Bu] and PKC-Cel that is equivalent to [Ch][Ac] images. Both ILs had distinct native EFB morphologies after IL-cellulase integration and pretreatment. Differences have been predicted to be observed due to the deterioration of lignin and decreased crystallinity of cellulose that interrupts the tissue system. Nonetheless, this has been discovered in both FTIR study and biomass chemical analysis. Figures 3.1b and 3.2b demonstrate how the surface is extended and disrupted in contrast to the native EFB. The relationship with the integrated framework became apparent. The native EFB surface was rough and solid, Figs. 3.1a and 3.1a but it gradually began to be damaged in the presence of the IL.

Morphology and characterization of surface modification confirmed that there was a compact immaterial pore structure in EFB-pretreated samples. As seen in Figs. 3.1c and 3.2c, the pretreated EFB had more open structures.

3.6.1.4 Fourier Transform Infrared Spectroscopy (FTIR)

Figure 3.3 displays both [Ch][Ac] and [Ch][Bu] spectrum of FTIR on native, hydrolyzed and IL-pretreated specimens. In [Ch][Ac] pretreated samples, the absorbance bands have a lower intensity range of approximately 1505–1557 cm^{-1}. 1503–1556 cm^{-1} was the result observed in [Ch][Bu]. Furthermore, pretreatment with ILs triggers the disruption of EFB lignin. The presence of cellulose was found

Fig. 3.1 SEM images of
a raw untreated EFB, **b** EFB
treated with [Ch][Bu], **c** EFB
treated with [Ch][Bu], at a
magnification of 1000X

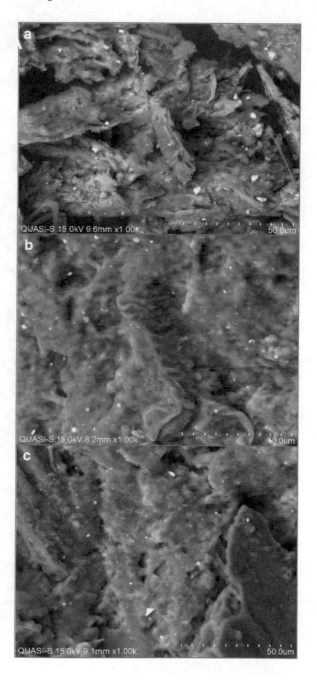

Fig. 3.2 SEM images of
a Raw untreated EFB, **b** EFB
treated with [Ch][Ac] and
PKC-Cel, **c** EFB treated with
[Ch][Ac] and PKC-Cel (after
one hour of the incubation
period), at a magnification of
1000X

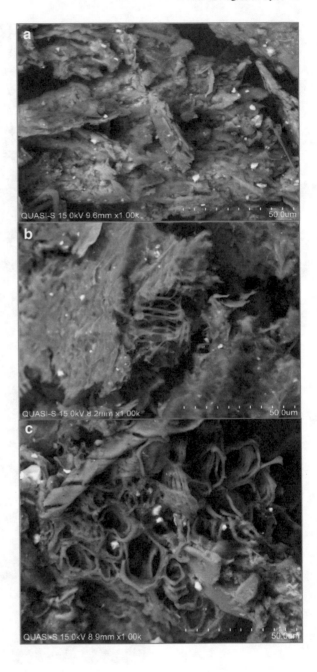

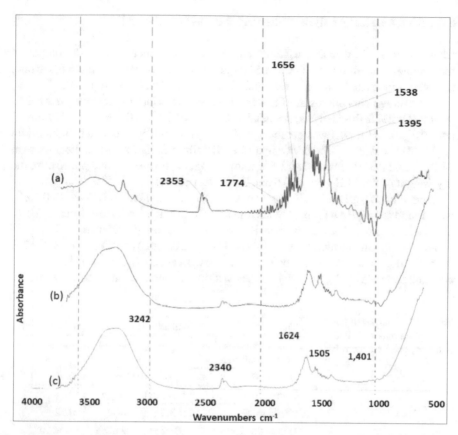

Fig. 3.3 FTIR spectra of a native (untreated) EFB at the absorbance bands of 500–4000 cm^{-1}, [b][Cho]Bu-pretreated EFB in the enzymatic system (IL-E), [c][Cho]OAc-pretreated EFB in the enzymatic system (IL-E) (Reprinted by permission from Springer Nature: Springer, 3 Biotech, Elgharbawy et al., 2018)

when the absorbance increased between 1430 and 896 cm^{-1} in contrast to native EFB (Nomanbhay et al., 2013).

Changes in [Ch][Ac] EFB treated sample were also observed at 1720–1740 cm^{-1} and a comparable range was observed in [Ch][Bu] EFB untreated sample. The decreasing spectra of sugar peaks ranging from 1707 to 1584 cm^{-1} and 2947 to 3633 cm^{-1} were integrated with PKC-Cel after [Ch][Ac] and [Ch][Bu] pretreatment. Cellulose II recorded the band at 895 cm^{-1}, which is highly acquiescent to hydrolysis.

3.6.1.5 Chemical Analysis of the Pretreated-Hydrolyzed EFB

Chemical analysis was also used after hydrolysis to gather data on the output of the sample, in addition to FTIR and SEM analysis. As shown in Table 3.5, carbon, nitrogen sources and minerals were included in the hydrolysate produced.

The average phosphorus in EFB was 11.8 mg/g contributing total weight of EFB around 0.98%. In this analysis, the metal ions Mg^{2+} (0.103 mg/L), Mn^{2+} (0.171 mg/L) and Zn^{2+} (0.183 mg/L) was obtained with 0.3 g/L of nitrogen content. Baharuddin et al. (2011) calculated the dry weight of EFB cellulose, lignin and hemicellulose at 50.3, 18 and 26.1%, respectively. For total nitrogen, phosphorus and potassium, the dry weight of the EFB were 1, 0.6 and 2.3%, respectively.

Producing 120 g/L of EtOH, Mg^{2+} (0.05 g/L), Mn^{2+} (0.04 g/L) and Zn^{2+} (0.01 g/L) were found as best EtOH supplements for sweet sorghum juice (Deesuth et al., 2012). The findings prove that they are consistent with the reported evidence.

According to a relevant study, furfural acid (2.1 ± 0.5 g/L) and CH_3COOH (0.62 ± 0.05 g/L) did not affect EtOH production (Ylitervo et al., 2013). The peak CH_3COOH concentration was 3.0 g/L, which comes within an adequate range for metabolism

Table 3.5 Analysis of the hydrolysate obtained after [Ch][Ac]-cellulase hydrolysis

Entry	Value/Unit	% (w/w) EFB
Minerals composition of the hydrolysate		
Calcium, Ca	0.6 ± 0.05 mg/g	0.06
Sodium, Na	1.670 ± 0.05 mg/g	0.14
Potassium, K	3.2 ± 0.5 mg/g	0.32
Copper, Cu	0.02 ± 0.007 mg/g	1.5×10^{-3}
Total Phosphorus, TP	11.8 mg/L	0.98
Iron, Fe	0.3 ± 0.01 ppm	1.40×10^{-4}
Zinc, Zn	0.2 ± 0.01 ppm	1.30×10^{-4}
Manganase, Mn	0.2 ± 0.01 ppm	1.24×10^{-4}
Magnesium, Mg	0.1 ± 0.01 ppm	1.70×10^{-5}
Chemical composition of the hydrolysate		
Protein	236.7 ± 2.75 μg/mL	1.27
Total soluble sugars, TOS	4.7 ± 0.7 mg/mL	0.47
Total reducing sugar, TRS	79.0 ± 1.65 g/L	61.9
Total Nitrogen, TN	300.0 ± 6.06 mg/L	2.12
Ttotal phenolic compounds, TPC	0.4 ± 0.05 mg/g	0.04
Acetic acid	0.6 ± 0.05 g/L	0.03
Furfural	2.0 ± 0.5 g/L	0.11

inhibition (Palmqvist & Hahn-Hägerdal, 2000). The total phenolic compounds (0.36 0.05 mg/g) did not fall in the fermentation suppression range. Free amino nitrogen may be replaced by an internal source of nitrogen through hydrolysis. The study showed that total nitrogen (TN) was found to be 0.237 g/L (1.27% w/w), while total protein (TP) is 0.3 g/L (2.12% w/w). Despite the low concentrations, the information reported on EFB is still comparable, showing that dry protein weight was approximately 2.3 ± 0.1%.

In Fig. 3.4, glucose (45.47%) was the most prevalent component in the mixture followed by xylose (19.65%) after investigated by HPLC in the hydrolysate monosaccharides of sugar portion. Low sugars presented in the hydrolysate included mannose (0.45%), galactose (0.16%) and arabinose (0.087%). Galactose and arabinose contributed to the lowest levels attained in this term.

EFB generated 22.0 ± 0.4% (w/w) lignin insoluble, 24.2 ± 0.5% (w/w) MGX (mannan, galactan and xylan), 36.6 ± 0.6% (w/w) glucan and 1.2 ± 0.1% (w/w) arabin when tested for carbohydrate content. The remaining was produced in the form of crude protein and extractive (Kim et al., 2015). The findings of further research are compatible with the most recent published studies.

3.6.1.6 Swelling Capacity and Crystallinity Index (CrI)

As seen in Table 3.6, Crystallinity (CrI), [Ch][Bu] was more effective than [Ch][Ac] at reducing cellulose crystallinity, decreasing the CrI value by integrating both IL and PKC-Cel into a single framework. [Ch][Bu], in addition to [Ch][Ac], however has a complex thermal expansion and standard enthalpy. This indicates [Ch][Ac] has higher molecular energy (Muhammad et al., 2012).

Figure 3.5 demonstrate maximum cellulase adsorption was detected in EFB, followed by (79.3 ± 2.8) g/L [Ch][Ac] and (81.4 ± 3.3) g/L [Ch][Bu]. The IL pretreatment can soften the compounds and extract them from EFB. In contrast to the crude samples, the desorption rate of cellulase increased in the IL-E treated samples.

3.6.1.7 Fermentation of Hydrolysate to Ethanol

After 48 h of 0.62 ± 0.06 g/g EFB hydrolysis, the average content of sugar is estimated around 70–75 g/L. For fermentation media, EtOH production was used for hydrolysate. Figure 3.6 demonstrates the profile concentration of ethanol and sugar over 84 h duration of the fermentation period. When a higher number of cells produces, so does the amount of ethanol. Production of EtOH hit its peak at 72 h and then decreased after 72 h. After 48 h, a sudden decline in sugar concentration was detected, while EtOH had begun to rise drastically. The EtOH yield was obtained around 0.4178 g/L/h. As observed in the recent study, the production of EtOH was 0.275 g/g of EFB. Without additional supplement after 72 h, 34.58 g/L of EtOH output was obtained.

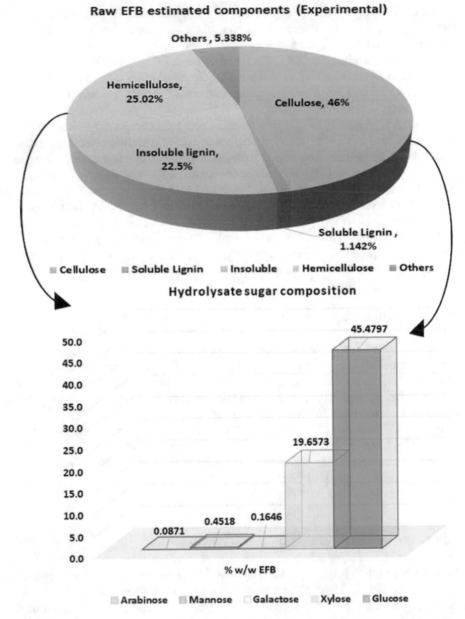

Fig. 3.4 Detailed analysis of sugar components for raw EFB and its hydrolysate obtained after treatment in the IL-E system. Pretreatment time and temperature: 60 min and 75 °C, hydrolysis time: 48 h, IL concentration 10%, PKC-Cel loading 45 FPU/g EFB, and temperature 50 °C (Reprinted by permission from Springer Nature: Springer, 3 Biotech, Elgharbawy et al., 2018)

Table 3.6 Analysis of the treated and treated hydrolyzed EFB

Entry	Crystalinity index (CrI)	Swelling capacity (g/g)
Native EFB	2.1 ± 0.55	0.4 ± 0.32
Treated EFB, [Ch][Bu]	1.3 ± 0.45	1.7 ± 0.44
Treated EFB, [Ch][Ac]	1.6 ± 0.65	1.8 ± 0.56
Treated EFB, [Ch][Bu] + E	0.6 ± 0.05	NA[a]
Treated EFB, [Ch][Ac] + E	0.6 ± 0.05	NA[a]

[a]Not applicable: as the mixture, in this case, was not in solid form due to performing one-step pretreatment and hydrolysis Crystallinity index = A1430/A896

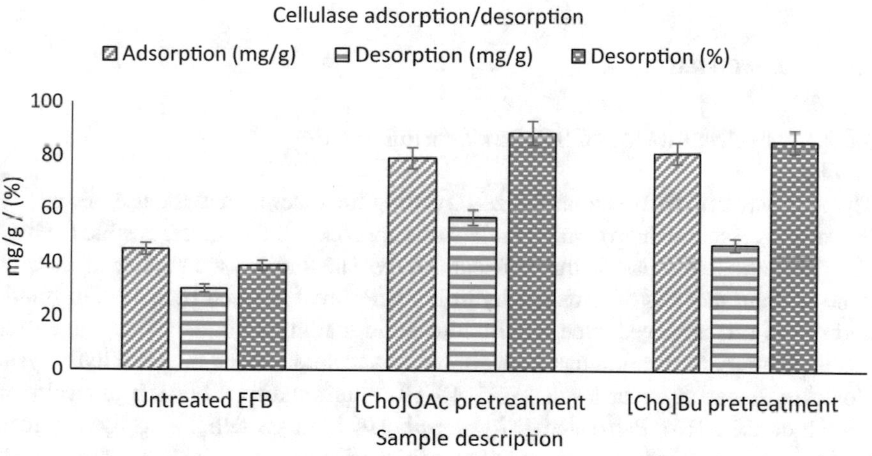

Fig. 3.5 Cellulase adsorption and desorption on EFB treated with [Ch][Ac] and [Cho][Bu]. Data include treated samples and raw samples

The EtOH yield observed was 0.127 g EtOH/g EFB for the raw EFB, while 0.275 g EtOH/g EFB was produced by the IL-E system; 87.94% of the theoretical output. In different pretreatment conditions, the findings achieved in the present research were contrasted with the reported data.

For pretreatment with *Eucalyptus*, [EMIM][Ac] was used, which was then given 0.172 g EtOH/g biomass after Novo-Celluclast hydrolyzed (Lienqueo et al., 2016). Without supplementation, results showed the ability to transform EFB hydrolysate to EtOH.

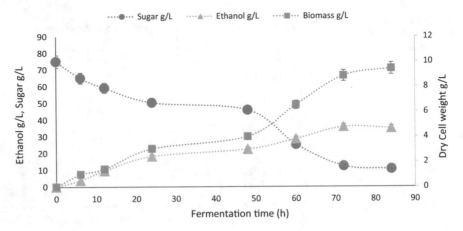

Fig. 3.6 Yeast growth, sugar consumption, and ethanol production during fermentation of IL-EFB hydrolysate and raw EFB samples for bioethanol production

3.6.2 Discussion

3.6.2.1 Pretreatment and Saccharification of EFB

The ability to recycle and immobilize IL during the reaction and the use of reaction in a single vessel can help commercialize this approach in the future, despite the high IL's high ratio to biomass in this lab-scale study. The findings are similar to research done without any ILs on biomass hydrolysis. Pretreated weed biomass enzymatic hydrolysis was achieved under continuous solid loading with mechanical agitation and sonication. Sonication has been shown to accelerate weed biomass hydrolysis. However, lowering sugar levels to 35–40 g/L required a total of 10 h of treatment (Borah et al., 2016). Pretreated *P. hysterophorus* biomass delignified by sulphuric acid was enzymatically hydrolyzed, then achieved approximately 400 mg of sugar/g of substrate. Considering the shorter time and enzyme intake, the findings obtained in the current study (EFB 600 mg/g) were higher in both cases (Singh et al., 2015).

Saccharification was supported in part by the distribution of lignin and hemicellulose. ILs anions act as hydrogen bond acceptors by interfering and degrading the hydroxyl group's structure in cellulose. Meanwhile, the cations interact with lignin and also with the interactions of π-π with hydrogen bonding (Asakawa et al., 2015). The inhibitory effect of PKC-Cel between lignocellulose and component of IL should be minimized as a result of these interactions. Zhao (2016) investigated whether hydrate cosmotropic occurs when enzymes are inactivated due to a lack of water or anions borderline (such as chloride and acetate) in high H-bond basic ILs. As a result, in low-water ILs (1%), cosmotropic anion activating enzymes like Cl^- and Ac^- in diluted aqueous solutions to become enzyme-inactivating agents. Perhaps this can explain PKC-resistance cellulase during the hydrolysis process to the IL-enzyme solution.

Previous studies by Elgharbawy et al. (2016a) found the highest production of sugar with [Ch][Ac]. The study yielded samples treated with [Ch][Ac] containing 22.5 g/L xylose, 0.09 g/L arabinose and 24 g/L glucose. Untreated EFB hydrolysis released 0.50 g/L (xylose), 0.13 g/L (arabinose) and 0.12 g/L (glucose) under the same conditions of 45 FPU/g EFB, at 50 °C for 24 h of enzyme hydrolysis without IL pretreatment. The sugar test shows that hydrolysate contains a number of sugars, including glucose, fructose, galactose, ribose and glyceraldehyde.

IL anions and hydroxyl hydrogen particles interact in the cellulose chain by forming hydrogen bonds. The hydrolytic rate rises and more sugar is produced as these bonds interact with the cellulose H-bonding mechanism. Furthermore, owing to IL application, the degree of crystallinity and polymerization decreases, hence weakening the composition of biomass (Yang & Fang, 2015).

Because of EFB pore enlargement and lignin disruption, IL's effect was three times greater than native samples due to the increase of cellulose hydrolysis (Sun et al., 2013). The structure of cellulose was subjected to polymorphic modifications during and after treatment. The structural improvements made it possible for EFB to be hydrolyzed by enzymes. Furthermore, cellulose I is a hydrolysis-resistant crystalline form. Enzymatic hydrolysis would be more susceptible to cellulose I conversion to other types of cellulose (amorphous structures or cellulose II) (Samayam et al., 2011).

3.6.2.2   Changes in Cellulose, Hemicellulose, and Lignin

In this analysis, the ability of ILs to interrupt the development of lignocellulosic biomass was demonstrated. Pretreatment of ILs affects hydrolysis, which affects the production of sugar and glucose. The content of the cellulose evaluated differed. The elimination of lignin triggers the hydrolysis of biomass, and the pretreatment with ILs carries out a significant and important task. The elimination of lignin makes the hydrolysis stage easier.

3.6.2.3   Scanning Electron Microscopy (SEM) for Morphological Structural Changes

The IL and PKC-Cel pretreatments resulted in a smooth and swollen structure and surface changes that exposed the cellulose fiber in one process. In general, as seen in the following sections, lignocelluloses with the fewest structure crystalline are ideally suited for hydrolysis by the enzyme.

3.6.2.4   Fourier Transform Infrared Spectroscopy (FTIR)

EFB treatment with hemicellulose, the structure of AFEX and lignin was reported to be between 1800 and 900 cm^{-1} in a report published by Abdul et al. (2016).

The peak modifications were predicted by the aromatic ring of lignin (C = C vibration). Particularly, EFB lignin degradation is facilitated by [Ch][Bu] and [Ch][Ac]. Cellulose (C-H bonding) and hemicellulose (C-O-C band vibration) were identified as causes of reduced peak intensity. Furthermore, EFB absorbance ($1738 \, \text{cm}^{-1}$) in hemicelluloses is related to carbonyl and carboxylic bonds (Abdul et al., 2016).

3.6.2.5 Chemical Analysis of the Treated-Hydrolyzed EFB

It is worth noting that the yeast's ability to produce EtOH is dependent on a range of factors, including nutrients and microbial strains. Carbon and nitrogen are important supplements to the media. Nitrogen is essential for the growth of yeast and affects ethanol generation as well as tolerance of EtOH. Micronutrients, in addition to nitrogen and carbon, are needed for cell growth and fermentation.

Zn^{2+}, Mg^{2+}, and Mn^{2+} are essential trace elements for the development of yeast and the processing of ethanol. Zinc influences cell growth as well as yeast cell metabolism. The Zn^{2+}-supplemented culture had improved the production and resistance of ethanol. Magnesium is necessary for yeast cell formation and cell production and enzymatic activity as a cofactor enzyme. It has a beneficial effect on EtOH production as it lowers proton by stabilizing the bilayer and interacting with the phospholipid layer for anion to enter the plasma layer. This promotes an enhancement in the resistance of ethanol yeast. The Mn^{2+} metal ion is a crucial element in the metabolism of *S. cerevisiae*, since it is a part of some proteins that make fermentation-related enzymes like carboxylase pyruvate (Deesuth et al., 2012).

During pretreatment and hydrolysis, microorganism-inhibiting compounds such as CH_3COOH, carbohydrate-derived compounds, and lignin by-products such as furfural and phenolic acids may form.

CH_3COOH may be released into the medium after solubilization and hemicellulose hydrolysis. Furfural is formed as xylose is degraded in acidic environments, but comparatively high amounts can be obtained by thorough pretreatment with lignocellulose (Palmqvist et al., 1999).

Furfural reduction may result in its respective alcohol being catalyzed by alcohol dehydrogenase, which has a less inhibitory impact (Dabirmanesh et al., 2012). By reducing the effect of inhibition, high cell concentrations will lead to faster yeast cell bioconversion in the presence of furfurals. Without significant changes in ethanol processing, *saccharomyces cerevisiae* can tolerate elevated furfural concentrations at elevated cell concentrations. The yeast successfully released ethanol in a culture medium containing 17.0 g/L furfural (Ylitervo et al., 2013).

In the current research, much of the cellulose has been broken down into glucose (99%). The enzymatic method showed the efficacy of hydrolysis based on the analysis. The transition rate from hemicellulose to xylose was 78.8%, with the rest being arabinose, galactose, and mannose. The total sugar produced may result from unreacted sugars or monosaccharides which have not been identified by the HPLC.

The presence of mannose, xylose, arabinose, and galactose in the mixture fit the composition of the EFB as reported. The components obtained were determined to

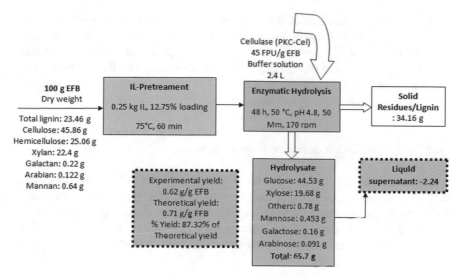

Fig. 3.7 Detailed biomass balance of the EFB pretreatment and hydrolysis in IL-E system. The process starts with 1.875 g of the EFB (*Source* Elgharbawy et al., 2018)

EFB initial used, as shown in Fig. 3.7. In hydrolysate, several EFBs were broken down into xylose and glucose, yielding 1.23 g of sugar. With just 0.5% error, the experimental yield for the conversion of cellulose to glucose was 87.3% of the theoretical yield.

3.6.2.6 Swelling Capacity and Crystallinity Index (CrI)

Pretreatment at low temperatures allowed the structure to swell and cellulose crystallinity to decrease, while high-temperature pretreatment produced large pores and disrupted the biomass (Noori & Karimi, 2016). IL can degrade the hydrogen bonds in the cellulose system, allowing enzymes to hydrolyze cellulose I into cellulose II. IL accessing the amorphous area of the cellulose substance induces swelling, which weakens hydrogen bonds, increases the portability of the cellulose chain, and decreases crystalline components. The greater the EFB swelling, the greater availability that enables hydrolysis.

Cations enter and pass through tiny pores in cellulose during IL pretreatment, improving the pretreated lignocellulose polyionic character. The cation can acts as a counter-charge to COO in structure, causing the material to swell (Reddy, 2015). Reducing the crystallinity of cellulose dissolution is a perfect way to enhance IL enzymatic hydrolysis at low temperatures (Xu et al., 2017). The primary effect of IL on lignocellulose at high temperatures is destruction in the structure which contributes to the decomposition of hemicellulose and lignin (Soudham et al., 2015).

IL pretreatment increased cellulase desorption by improving the surface area and reducing the content of lignin, resulting in successful hydrolysis. The tendency of lignin to irreversibly absorb cellulase is one of the potential causes of cellulase's deactivation. In the presence of IL, hydrolysis will also decrease the content of lignin because it can bind to IL, exposing cellulose that can facilitate cellulose desorption while also increasing hydrolysis (Zhao, 2016).

3.6.2.7 Fermentation of Hydrolysate to Ethanol

Fermentation has not hindered the development of yeast cells in the presence of IL. The graph also shows that as the incubation duration grew, the optical density corresponding to the output increased as well. Choline has been reported as a vitamin B compound and is commonly used as a dietary supplement; ILs that rely on it are known to be environmentally safe and thus considered non-toxic (Ossowicz et al., 2016).

3.7 Conclusion

In the IL-E process, an integrated IL ([Ch][Ac]) of EFB pretreatment and PKC-Cel (cellulase) yielding theoretical glucose output of around 87.3%. After 48 h in the [Ch][Ac]-enzyme method, about 99% of glucose was converted from cellulose. The hydrolysate of EFB was evaluated as a potential EtOH refining process. Furthermore, IL's presence in the fermentation medium had no detrimental effect on yeast cell growth due to vitamin B compound containing in IL.

Acknowledgements The project was funded by the Ministry of Higher Education (MOHE) in Malaysia under the FRGS-13-088-0329 research grant. We thank the Research Management Center, IIUM for the grant [RMCG20-021-0021RMCG20-021-0021]. We also thank the International Institute for Halal Research and Training (INHART), IIUM Department of Biotechnology Engineering and Ionic Liquid Research Center (CORIL), UTP for assisting us with the essential laboratory equipment. We are also thankful for the raw materials provided by Sime Darby Plantation (West Oil Mill).

References

Abdul, P. M., Jamaliah, M. J., Suhaida, H., Masturah, M., Nabilah, L., Hassan, O., Venkatesh, B., Dale, B. E., & Mohd, T. M. N. (2016). Effects of changes in chemical and structural characteristic of ammonia fibre expansion (AFEX) pretreated oil palm empty fruit bunch fibre on enzymatic saccharification and fermentability for biohydrogen. *Journal of Bioresource Technology, 211,* 200–208.

Asakawa, A., Kohara, M., Sasaki, C., Asada, C., & Nakamura, Y. (2015). Comparison of choline acetate ionic liquid pretreatment with various pretreatments for enhancing the enzymatic saccharification of sugarcane bagasse. *Industrial Crops and Products, 71,* 147–152. https://doi.org/10.1016/j.indcrop.2015.03.073.

Baharuddin, A. S., Rahman, N. A., Shan, U. K., Hassan, M. A., Wakisaka, M., & Shirai, Y. (2011). Evaluation of pressed shredded empty fruit bunch (EFB)-palm oil mill effluent (POME) anaerobic sludge based compost using Fourier transform infrared (FTIR) and nuclear magnetic resonance (NMR) analysis. *African Journal Biotechnology, 10,* 8082–8289.

Bian, J., Peng, F., Peng, X., Xiao, X., Peng, P., & Xu, F. (2014). Effect of [Emim] Ac pretreatment on the structure and enzymatic hydrolysis of sugarcane bagasse cellulose. *Carbohydrate Polymers, 100,* 211–217. https://doi.org/10.1016/j.carbpol.2013.02.059.

Borah, A. J., Agarwal, M., Poudyal, M., Goyal, A., & Moholkar, V. S. (2016). Mechanistic investigation in ultrasound induced enhancement of enzymatic hydrolysis of invasive biomass species. *Bioresource Technology, 213,* 342–349. https://doi.org/10.1016/j.biortech.2016.02.024.

Caputi, A., Ueda, M., & Brown, T. (1968). Spectrophotometric determination of ethanol in wine. *American Journal of Enology and Viticulture, 19,* 160–165.

Chandra, R., Takeuchi, H., & Hasegawa, T. (2012). Methane production from lignocellulosic agricultural crop wastes: A review in context to second generation of biofuel production. *Renewable and Sustainable Energy Reviews, 16*(3), 1462–1476. https://doi.org/10.1016/j.rser.2011.11.035.

Chew, T. L., & Bhatia, S. (2008). Catalytic processes towards the production of biofuels in a palm oil and oil palm biomass-based biorefinery. *Bioresource Technology, 99,* 7911–7922. https://doi.org/10.1016/j.biortech.2008.03.009.

Choi, W. I., Park, J. Y., Lee, J. P., Oh, Y. K., Park, Y. C., Kim, J. S., Park, J. M., Kim, C. H., & Lee, J. S. (2013). Optimization of NaOH catalyzed steam pretreatment of empty fruit bunch. *Biotechnology for Biofuels, 6,* 170.

Chong, P. S., Jahim, J. M., Harun, S., Lim, S. S., Sahilah, A. M., Osman, H., & Mohd, T. M. N. (2013). Enhancement of batch biohydrogen production from prehydrolysate of acid treated oil palm empty fruit bunch. *International Journal of Hydrogen Energy, 38,* 9592–9599.

Dabirmanesh, B., Khajeh, K., Ranjbar, B., Farideh, G., & Heydari, A. (2012). Inhibition mediated stabilization effect of imidazolium based ionic liquids on alcohol dehydrogenase. *Journal of Molecular Liquids, 170,* 66–71. https://doi.org/10.1016/j.molliq.2012.03.004.

Deesuth, O., Laopaiboon, P., Jaisil, P., & Laopaiboon, L. (2012). Optimization of nitrogen and metal ions supplementation for very high gravity bioethanol fermentation from sweet sorghum juice using an orthogonal array design. *Energies, 5,* 3178–3197.

Dubois, M., Gilles, K. A., Hamilton, J. K., Hamilton, P. A., Reber, P. A., & Fred, S. (1956). Colorimetric method for determination of sugar and related substances. *Analytical Chemistry, 28,* 350–356. https://doi.org/10.1021/ac60111a017.

Elgharbawy, A. A., Alam, M. Z., Kabbashi, N. A., Moniruzzaman, M., & Jamal, P. (2016a). Implementation of definite screening design in optimization of in situ hydrolysis of EFB in cholinium acetate and locally produced cellulose combined system. *Waste Biomass Valorization.* https://doi.org/10.1007/s12649-016-9638-6 (LB—Elgharbawy 2016).

Elgharbawy, A. A., Alam, M. Z., Kabbashi, N. A., Moniruzzaman, M., & Jamal, P. (2016b). Evaluation of several ionic liquids for in situ hydrolysis of empty fruit bunches by locally-produced cellulase. *3 Biotech, 6,* 128. https://doi.org/10.1007/s13205-016-0440-8.

Elgharbawy, A. A., Alam, Z., Moniruzzaman, M., & Goto, M. (2016c). Ionic liquid pretreatment as emerging approaches for enhanced enzymatic hydrolysis of lignocellulosic biomass. *Biochemical Engineering Journal, 109,* 252–267. https://doi.org/10.1016/j.bej.2016.01.021.

Elgharbawy, A. A., Alam, M. Z., Moniruzzaman, M., Kabbashi, N. A., & Jamal, P. (2018). Chemical and structural changes of pretreated empty fruit bunch (EFB) in ionic liquid-cellulase compatible system for fermentability to bioethanol. *3 Biotech, 8*(5), 236. https://doi.org/10.1007/s13205-018-1253-8.

Ghose, T. K. (1987). Measurement of cellulase activities. *International Union of Pure and Applied Chemistry, 59,* 257–268. https://doi.org/10.1351/pac198759020257.

Graenacher, C. (1934). Patent No. 1,943,176.

Hahn-Hägerdal, B., Galbe, M., Gorwa-Grauslund, M. F., Liden, G., & Zacchi, G. (2006). Bioethanol: The fuel of tomorrow from the residues of today. *Trends in Biotechnology, 24,* 549–556.

Ishola, M. M., Isroi, & Taherzadeh, M. J. (2014). Effect of fungal and phosphoric acid pretreatment on ethanol production from oil palm empty fruit bunches (OPEFB). *Bioresource Technology, 165,* 9–12. https://doi.org/10.1016/j.biortech.2014.02.053.

Kim, D. Y., Um, B. H., & Oh, K. K. (2015). Acetic acid-assisted hydrothermal fractionation of empty fruit bunches for high hemicellulosic sugar recovery with low byproducts. *Applied Biochemistry and Biotechnology, 176,* 1445–1458.

Kljun, A., Benians, T. A. S., Goubet, F., Meulewaeter, F., Knox, J. P., & Blackburn, R. S. (2011). Comparative analysis of crystallinity changes in cellulose I polymers using ATR-FTIR, X-ray diffraction, and carbohydrate-binding module probes. *Biomacromolecules, 12,* 4121–4126. https://doi.org/10.1021/bm201176m.

Laureano-Perez, L., Teymouri, F., Alizadeh, H., & Dale, B. E. (2005). Understanding factors that limit enzymatic hydrolysis of biomass. *Applied Biochemistry and Biotechnology, 124,* 1081–1099. https://doi.org/10.1385/abab:124:1-3:1081.

Lienqueo, M. E., Ravanal, M. C., Pezoa-Conte, R., Cortinez, V., Martinez, L., Niklitschek, T., Salazar, O., Carmona, R., Garcia, A., Hyvarinen, S., Maki-Arvela, P., & Mikkola, J.-P. (2016). Second generation bioethanol from eucalyptus globulus Labill and Nothofagus pumilio: Ionic liquid pretreatment boosts the yields. *Industrial Crops and Products, 80,* 148–155.

MacFarlane, D. R., Kar, M., & Pringle, J. M. (2017). An introduction to ionic liquids. In *Fundamentals of ionic liquids: From chemistry to application* (pp. 1–25).

Moniruzzaman, M., & Goto, M. (2018). Ionic liquid pretreatment of lignocellulosic biomass for enhanced enzymatic delignification. In: Itoh T, & Koo Y-M (Eds.), *Application of ionic liquids in Biotechnology,* (Vol. 168, pp. 61–77). Springer International Publishing, Cham. https://doi.org/10.1007/10_2018_64. PMID: 29744542.

Moniruzzaman, M., Nakashima, K., Kamiya, N., & Goto, M. (2010). Recent advances of enzymatic reactions in ionic liquids. *Biochemical Engineering Journal, 48,* 295–314. https://doi.org/10.1016/j.bej.2009.10.002.

Muhammad, N., Hossain, M, I., Man, Z., El-Harbawi, M., Bustam, M. A., Noaman, Y. A., Alitheen, N. B. M., Ng, M. K., Hefter, G., & Yin, C.-Y. (2012). Synthesis and physical properties of choline carboxylate ionic liquids. *Journal of Chemical & Engineering Data, 57,* 2191–2196.

Nieves, D. C., Ruiz, H., De Cárdenas, L. Z., Alvarez, G. M., Aguilar, C. N., Ilyina, A., & Hernandez, J. L. M. (2016). Enzymatic hydrolysis of chemically pretreated mango stem bark residues at high solid loading. *Industrial Crops and Product, 83,* 500–508. https://doi.org/10.1016/j.indcrop.2015.12.079.

Ninomiya, K., Kohori, A., Tatsumi, M., Osawa, K., Endo, T., Kakuchi, R., Ogino, C., Shimizu, N., & Takahashi, K. (2015a). Ionic liquid/ultrasound pretreatment and in situ enzymatic saccharification of bagasse using biocompatible cholinium ionic liquid. *Bioresource Technology, 176,* 169–174. https://doi.org/10.1016/j.biortech.2014.11.038.

Ninomiya, K., Ogino, C., Ishizaki, M., Yasuda, M., Shimizu, N., & Takahashi, K. (2015b). Effect of post-pretreatment washing on saccharification and co-fermentation from bagasse pretreated with biocompatible cholinium ionic liquid. *Biochemical Engineering Journal, 103,* 198–204. https://doi.org/10.1016/j.bej.2015.08.002.

Ninomiya, K., Omote, S., Ogino, C., Kuroda, K., Noguchi, M., Endo, T., Kakuchi, R., Shimizu, N., & Takahashi, K. (2015c). Saccharification and ethanol fermentation from cholinium ionic liquid-pretreated bagasse with a different number of post-pretreatment washings. *Bioresource Technology, 189,* 203–209. https://doi.org/10.1016/j.biotech.2015.04.022.

Nomanbhay, S. M., Hussain, R., & Palanisamy, K. (2013). Microwave-assisted alkaline pretreatment and microwave assisted enzymatic saccharification of oil palm empty fruit bunch fiber for enhanced fermentable sugar yield. *Journal of Sustainable Bioenergy Systems, 3,* 7–17.

Noori, M. S., & Karimi, K. (2016). Detailed study of efficient ethanol production from elmwood by alkali pretreatment. *Biochemical Engineering Journal, 105,* 197–204. https://doi.org/10.1016/j.bej.2015.09.019.

Ossowicz, P., Janus, E., Szady-Chełmieniecka, A., & Rozwadowski, Z. (2016). Influence of modification of the amino acids ionic liquids on their physico-chemical properties: Ionic liquids versus ionic liquids supported Schiff bases. *Journal of Molecular Liqiud, 224,* 211–218. https://doi.org/10.1016/j.molliq.2016.09.111.

Palmqvist, E., Grage, H., Meinander, N. Q., & Hahn-Hägerdal, B. (1999). Main and interaction effects of acetic acid, furfural and p hydroxybenzoic acid on growth and ethanol productivity of yeast. *Biotechnology and Bioengineering, 63,* 46–55. https://doi.org/10.1002/(SICI)1097-0290(19990405)63.

Palmqvist, E., & Hahn-Hägerdal, B. (2000). Fermentation of lignocellulosic hydrolysates. II: inhibitors and mechanisms of inhibition. *Bioresource Technology, 74,* 25–33. https://doi.org/10.1016/s09608524(99)001613.

Poornejad, N., Karimi, K., & Behzad, T. (2014). Ionic liquid pretreatment of rice straw to enhance saccharification and bioethanol production. *Journal of Biomass to Biofuel, 1,* 8–15. https://doi.org/10.11159/jbb.2014.002.

Reddy, P. (2015). A critical review of ionic liquids for the pretreatment of lignocellulosic biomass. *South African Journal of Science, 111,* 1–9.

Salvador, Â. C., Santos, M. D. C., & Saraiva, J. A. (2010). Effect of the ionic liquid [bmim] Cl and high pressure on the activity of cellulase. *Green Chemistry, 12,* 632–635.

Samayam, I, P., Hanson, B. L., Langan, P., & Schall, C. A. (2011). Ionic-liquid induced changes in cellulose structure associated with enhanced biomass hydrolysis. *Biomacromolecules.* https://doi.org/10.1021/bm200736a (LB—Samayam 2011).

Shi, W,, Jia, J, Gao, Y., & Zhao, Y. (2013). Influence of ultrasonic pretreatment on the yield of bio-oil prepared by thermo-chemical conversion of rice husk in hot-compressed water. *Bioresource Technology, 146,* 355–362.

Singh, S., Agarwal, M., Bhatt, A., Goyal, A., & Moholkar, V, S. (2015). Ultrasound enhanced enzymatic hydrolysis of Parthenium hysterophorus: A mechanistic investigation. *Bioresource Technology, 192,* 636–645. https://doi.org/10.1016/j.biortech.2015.06.031.

Soudham, V, P., Raut, D, G., Anugwom, I., Brandberg, T., Larsson, C., & Mikkola, P. (2015). Coupled enzymatic hydrolysis and ethanol fermentation: Ionic liquid pretreatment for enhanced yields. *Biotechnology for Biofuels, 8,* 135. https://doi.org/10.1186/s13068-015-0310-3 (LB—Soudham 2015).

Sousa, L. D., Chundawat, S. P. S., Balan, V., & Dale, B. (2009). Cradle-to-grave assessment of existing lignocelluloses pretreatment technologies. *Current Opinion Biotechnology, 20,* 339–347.

Sun, N., Rahman, M., Qin, Y., & Maxim, M. L. (2009). Complete dissolution and partial delignification of wood in the ionic liquid 1-ethyl-3-methylimidazolium acetate †‡. *Green Chemistry, 11,* 646–655. https://doi.org/10.1039/b822702k.

Sun, Y.-C., Xu, J.-K., Xu, F., & Sun, R.-C. (2013). Structural comparison and enhanced enzymatic hydrolysis of eucalyptus cellulose via pretreatment with different ionic liquids and catalysts. *Process Biochemistry, 48,* 844–852. https://doi.org/10.1016/j.procbio.2013.03.023.

Updegraff, D. M. (1969). Semimicro determination of cellulose inbiological materials. *Analytical Biochemistry, 32,* 420–424. https://doi.org/10.1016/S0003-2697(69)80009-6.

Wang, X., Li, H., Cao, Y., & Tang, Q. (2011). Cellulose extraction from wood chip in an ionic liquid 1-allyl-3-methylimidazolium chloride (AmimCl). *Bioresource Technology, 102,* 7959–7965.

Wolfe, K., Wu, X., & Liu, R. H. (2003). Antioxidant activity of apple peels. *Journal of Agriculture and Food Chemistry, 51,* 609–614.

Xu, A.-R., Wen, S., & Chen, L. (2017). Dissolution performance of cellulose in MIM plus tetrabutylammonium propionate solvent. *Journal of Molecular Liquids, 246,* 153–156. https://doi.org/10.1016/j.molliq.2017.09.065.

Yang, C.-Y., & Fang, T. J. (2015). Kinetics of enzymatic hydrolysis of rice straw by the pretreatment with a bio-based basic ionic liquid under ultrasound. *Process Biochemistry, 50,* 623–629. https://doi.org/10.1016/j.procbio.2015.01.013.

Ylitervo, P., Franzén, C, J., & Taherzadeh, M, J. (2013). Impact of furfural on rapid ethanol production using a membrane bioreactor. *Energies, 6,* 1604–1617. https://doi.org/10.3390/en6031604.

Zainan, N. H., Alam, Z., & Al-Khatib, M. F. (2013). Production of sugar by hydrolysis of empty fruit bunches using palm oil mill effluent (POME) based cellulases: Optimization study. *African Journal of Biotechnology, 10,* 18722–18727.

Zarei, A. R. (2009). Spectrophotometric determination of trace amounts of furfural in water samples after mixed micelle-mediated extraction. *Acta Chimica Slovenica, 56,* 322–328.

Zhao, H. (2016). Protein stabilization and enzyme activation in ionic liquids: Specific ion effects. *Journal of Chemical Technology & Biotechnology, 91,* 25–50. https://doi.org/10.1002/jctb.4837.

Zhao, H., Jones, C. L., Baker, G. A., Xia, S., Olubajo, O., Person, V. N., & Ridge, O. (2009). Regenerating cellulose from ionic liquids for an accelerated enzymatic hydrolysis. *Journal of Biotechnology, 139,* 47–54. https://doi.org/10.1016/j.jbiotec.2008.08.009.

Chapter 4
Proliferation of Rat Amniotic Stem Cell (AFSC) on Modified Surface Microcarrier

Nurhusna Samsudin, Yumi Zuhanis Has-Yun Hashim, Hamzah Mohd Salleh, and Azmir Ariffin

Abstract Traditionally, stem cells are grown in two-dimensional vessels, although this poses a major challenge in the clinical-grade production of a large number of stem cells and their derivatives to fulfill the dose requirements in clinical trials (at a scale of $\sim 10^9 - 10^{10}$ cells). The cultivation of microcarriers in stirred bioreactors has been recognized as among the most promising strategies for the rapid scale-up of stem cell-derived therapeutics under controlled conditions. In addition to large-scale production, microcarriers have been applied in cell delivery systems for in vivo transplantation, to enhance cell survival and engraftment. 'Microcarrier' is a term used to refer to microspheres that support cells in mammalian cell culture, in which cells grow as monolayers on the surface of the particles. Microcarriers are spherical particles with a size ranging between 100 and 200 μm. Due to their small size, microcarriers have a wide variety of applications, one of which is cell culture and tissue engineering. This study proposes a novel model system for the in vitro study of cell proliferation ability on microcarriers.

Keywords Microcarrier · Stem cell · Pluripotent · Embryonic body

4.1 Introduction

Microspheres are spherical particles between 0.1 and 200 μm in size (Sahil et al., 2011). Due to its small size, microspheres can be categorized as a microparticle, microcarrier, and microcapsule. It has a wide variety of applications, e.g. in drug delivery systems, cell culture and tissue engineering, protein immobilization, and gene delivery. In each application, there are different terms used to indicate the particular use of microspheres specifically. The term microparticle is commonly used

N. Samsudin (✉) · Y. Z. H.-Y. Hashim · H. M. Salleh
International Institute for Halal Research and Training (INHART), International Islamic University Malaysia, Kuala Lumpur, Malaysia
e-mail: nurhusna@iium.edu.my

A. Ariffin
Faculty of Engineering Technology, University Malaysia Pahang, Kuantan, Malaysia

in the drug delivery system where a consistent and predictable particle surface area is important (Nikam et al., 2012). The microsphere that can entrap cells in the inner compartment is called a microcapsule (Tan et al., 2010). Meanwhile, the microsphere is used to support cells in mammalian cell culture which grow as monolayers on the surface of the microsphere is called a microcarrier (Brun-Graeppi et al., 2011; Tan et al., 2010). In this article, the term microcarrier is used as it is applied to cultivate and propagate mammalian cells.

4.1.1   Principle

The microcarrier cell culture system served two significant purposes. First, mass production of certain bioproducts, such as recombinant proteins, hormones, and vaccines, whereby animal cells are routinely cultured in a bioreactor to meet industrial demand (van der Velden-de Groot, 1995) and in a clinical trial stage (Goh et al., 2013). The second purpose is to serve as the delivery of cultured cells, and the transplantation of biodegradable microcarriers loaded with cultured cells into the body (Seland et al., 2011). Therefore, materials with appropriate degradation rates are beneficial in the microcarrier cell culture system to minimize the effect of a toxic degradation product. Formerly, in cell culture work, anchorage-dependent cells are typically activated on the wall of roller bottles or unagitated vessels, such as tissue culture flasks (White & Ades, 1990). This system is perfectly suited for research and lab scale. Moving to the industrial scale, these systems were no longer relevant due to limitations in culture space, cell yield, control in culture condition, and sterility. Particularly, industrial production includes the production of large quantities of mammalian cells and their bioproduct. Microcarrier is one of the most established technological platforms for industrial production to increase productivity (Chu & Robinson, 2001). Microcarrier acts as a substrate for cells to attach and is cultured in suspension in a bioreactor. There are various types of bioreactors used by microcarriers to grow mammalian cells, such as a stirred tank and a fluidized bed bioreactor.

Cell culture using microcarrier beads as a three-dimensional (3D) substrate was first introduced by Wezel (1967). This 3D cell culture has undergone extensive modifications to optimize conditions for cell propagation and simulating in vitro and in vivo conditions (Overstreet et al., 2003). Under proper conditions, the cells attach and spread to the carrier beads and gradually expand into a confluent monolayer (van der Velden-de Groot, 1995).

Three different types of polycaprolactone (PCL) based microcarriers that have been developed in previous research (Samsudin et al., 2017) (gelatin-coated PCL microspheres, UV/O_3-treated PCL, and untreated PCL microcarrier) were further tested to support the attachment and growth of rat amniotic fluid stem cells (AFSC).

Table 4.1 List of basic media component for cultivation of AFSC and spontaneous differentiation of AFSC

Media component	Volume (ml)
1 × GMEM	50
2.3% sodium bicarbonate	18.1
1 mM L-glutamine	2.9
0.5 mM sodium pyruvate	2.9
1 × NEAA	5.9
0.1 mM β-mercaptoethanol	1.1

4.2 Objective of Experiment

This study was set to determine the biocompatibility of the newly developed microcarrier, as well as to investigate the ability of AFSCs to maintain pluripotency after being cultivated in microcarrier culture.

4.3 Methodology

4.3.1 Cell Line

The cell line was kindly provided by Dr. Norshariza Nordin, Genetic and Regenerative Medicine Research Centre, Department of Obstetrics and Gynecology, Faculty of Medicine and Health Sciences, University Putra Malaysia. Amniotic fluid was collected from time mated Sprague Dawley rats as described in Mun-Fun et al. (2015).

4.3.2 Media Preparation

Two types of media were prepared. Essential stem media (ESM) for the cultivation of the AFSC and spontaneous stem cell (EBM) differentiation in a biocompatibility study. The basic media component (Table 4.1) was prepared and kept at 4 °C until further used. Mixing was conducted under a sterile condition in a biosafety hood.

4.3.3 Embryonic Stem Media (ESM)

ESM was prepared by mixing basic media with 15% FBS and 10 ng/mL of rat LIF and made to a total volume of 50 mL. It is best to prepare the ES media fresh before use.

4.3.4 Embryoid Bodies Media (EBM)

EBM was prepared by combining the basic media with 15% FBS and made to a total volume of 50 mL. It is best to prepare the EB media fresh before use.

4.3.5 Cell Propagation in a 2D Cultured Flask

4.3.5.1 Thawing of Cryopreserved Cells

The method suggested by Freshney (2010) was closely followed. The DEMEM media supplemented by 15% FBS was prepared accordingly before the thawing process. A vial of a cryopreserved cell (1 mL) from $-150°$ condition was warmed in a 37 °C water bath. Once the cryopreserved media starts to melt, DMEM medium (1 mL) was added into the vial and the mixture was mixed using a micropipette. The mixture was then transferred into a 15 mL centrifuge tube and centrifuged at $800 \times$ g for 5 min at 25 °C. The supernatant was discarded and 1 mL of fresh media containing 10% FBS was used to resuspend the cell pellet. About 4 mL of fresh medium were added to the homogeneous cell and the cells were counted. Cells were made to 1×10^5 cells/mL concentration seeded into 25 cm^2 T-flask containing media with the volume that was summed up to a total of 5 mL. The flask containing cells was then incubated in the CO_2 incubator supplied with 5% CO_2 at 37 °C.

4.3.5.2 Cell Counting

The concentration of cells in the suspension was determined using a haemacytometer with the aid of trypan blue. About 20 mL of cell suspension were mixed with an equal volume of trypan blue dye. The microliters of the mixture were placed on the haemacytometer and allowed to spread by capillary action. Cells were counted under an inverted microscope and the concentration of cells (cells/mL) was calculated using Eq. (4.1).

$$c = \frac{n}{v} \qquad (4.1)$$

c = cell concentration (cells/mL)
n = number of cells
v = volume counted (mL).

Standard Heamacytometer used to have the depth of chamber of 1 mm and the area of the central grid is 1 mm^2. Therefore $v = 0.1$ mm^3. The formula then becomes

$$c = {}^n\!/1 \times 10^{-4}$$ (4.2)

or

$$c = n \times 10^4$$ (4.3)

If the cells were too concentrated, the cell suspension can be diluted, and the dilution factor was added to the calculation as follows:

$$c = n \times \text{dilution factor} \times 10^4$$ (4.4)

4.3.5.3   Sub-culture of Cells

Media supplemented with 10% FBS was prepared according to the number and size of the flask to be used. Once the cells reached confluency or the media was exhausted, the cells were sub-cultured using a method described by Butler (2004). The spent media from confluent monolayer cells was carefully removed from the flask and the surface of the flask was washed with 2 mL of PBS to remove remaining FBS. One mL of Accutase (an enzyme with proteolytic and collagenolytic activity for detachment of cells from the flask's surface) was added to the flask and incubated in a CO_2 incubator for 5 min for detachment. About 4 mL of culture with 10% FBS was added to the flask, 1 mL of culture with 10% FBS was added to the flask. Cell detachment was monitored at 5 min intervals with a microscope to ensure the cell has been detached. To stop the reaction of the proteolytic enzyme, 1 mL of culture with 10% FBS was added to the flask. The cell's mixture was then removed into a 15-mL tube and centrifuged at $800 \times g$ for 5 min. The supernatant was discarded and the pellet was resuspended in 5 mL of media. Cell concentration was determined and was made to 1×10^5 cells/mL concentration seeded into a T-flask containing media. The flask containing the cells was then incubated in a CO_2 incubator supplied with 5% CO_2 at 37 °C.

4.3.6   Microcarrier Spinner Vessel Culture

4.3.6.1   Microcarriers Preparation

PCL microcarriers were sterilized using 70% ethanol rather than the standard auto-claving method, since the heat may cause the biopolymer to melt. PCL microcarriers (3 g/mL) were washed in Ca^{2+}, Mg^{2+}-free PBS. Once settled, the supernatant was decanted and was replaced by 70% (v/v) ethanol in distilled water. Microcarriers were washed twice with ethanol solution and then incubated overnight in 70% (v/v) ethanol. The ethanol solution was removed and microcarriers were rinsed three times in sterile Ca^{2+} and Mg^{2+}-free PBS (50 mL/g microcarrier) and once in culture medium (20–50 mL/g microcarriers) before use.

4.3.6.2 Spinner Vessel Culture

Cells were grown using a 500 mL spinner vessel with 200 mL working volume. Before cell inoculation, the inner surface of the spinner vessel was coated with 5% silicon oil in ethyl acetate to prevent microcarriers from attaching to the inner surface. After sterilization by standard autoclaving, the vessel was transferred to a biosafety cabinet. The culture medium (150 mL) supplemented by the desired concentration of fetal bovine serum (FBS) was transferred aseptically into the spinner vessel. This was followed by inoculation of 30 mL of culture medium containing suspended microcarriers, and 20 mL of culture medium supplemented by the desired concentration of FBS containing cells at a concentration of 1.5×10^5 cells/mL. The spinner vessel was then transferred to the humidified CO_2 incubator and agitated at low speed (30 rpm) for the first two hours and then continued at the desired agitation. Cell sampling was taken at 8 h–interval to determine cell growth. The cell growth on the newly developed biodegradable microcarrier was observed and the growth kinetic was calculated based on

$$\mu = \ln \frac{(x_1 - x_2)}{(t_1 - t_2)} \tag{4.5}$$

4.3.6.3 Sampling and Cell Counting

One mL of microcarriers culture was aseptically pipetted out from the spinner flask culture and placed in a 15 mL tube. The microcarriers could settle and the supernatant was discarded. Microcarriers were washed twice with PBS before treatment with Accutase and the tube was incubated in a CO_2 incubator for 15 min at 37 °C. After 15 min the mixture was gently flushed to detach the immobilized cells. The concentration of cells in the suspension was determined using a haemacytometer with the aid of trypan blue. Twenty mL of cell suspension was mixed with an equal volume of trypan blue dye. Ten microlitres of the mixture were placed on the haemacytometer and allowed to spread by capillary action. Cells were counted under an inverted microscope and the concentration of cells (cells/mL) was calculated using Eq. (4.2).

4.4 Result and Discussion

This study investigated the ability of the developed microcarriers to support rat amniotic fluid stem cell (AFSC), a primary cell line. AFSC was first isolated in 2007 by De Coppi et al. (2007) (as cited in Mun-Fun et al., 2015, p. 89) with high differentiation capacities that enable cells to develop into three primary germ layers linages (ectoderm, mesoderm, and endoderm). Under proper conditions, cells can be differentiated into dopaminergic neuron using a directed monolayer differentiation protocol which is potentially useful for Parkinson's disease treatment (Mun-Fun et al., 2015)

In this study, several AFSC responses upon interaction with microcarriers were determined. This includes AFSC adherence and proliferation during expansion on microcarriers, as well as their ability to retain cell shape and organization through enzymatic retrieval from microcarriers. The ability of the cells to retain their properties is a crucial aspect in maintaining the differentiation potential of the stem cell. The expansion of AFSCs in 3D cultures has been previously studied by Liu (2004). The result shows an improvement in cell yield when a microcarrier-based spinner flask culture system was used to scale up AFSC cell expansion.

Figure 4.1 shows the growth performance of AFSC on gelatin-coated PCL microcarrier, UV/O$_3$-treated PCL microcarrier, and untreated PCL microcarrier (control). Meanwhile, Table 4.2 shows the results of the calculated maximum cell number, growth rate, and the doubling time of the three microcarriers used. All experiments were carried out at a seeding concentration of 1.5×10^5 cells/mL and microcarrier concentration of 3 g/L. The number of cells that adhered to the microcarrier was

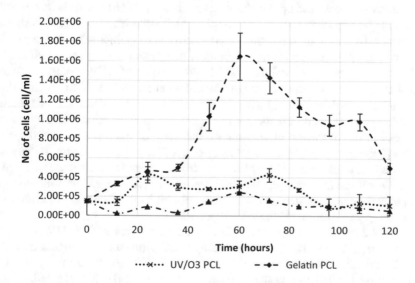

Fig. 4.1 Growth kinetics of rat amniotic stem cell (AFSC) on different microcarriers in stirred spinner flasks: (x) UV/O$_3$ PCL, (●) gelatin immobilized, (▲) untreated PCL

Table 4.2 Values of maximum cell concentration, growth kinetics and doubling time of AFSC cells on different types of microcarrier

Microcarrier	Maximum cell concentration ($\times 10^5$ cells/ml)	Growth rate, μ (h^{-1})	Doubling time, t_d (h)
Untreated PCL	2.4 ± 5.9	0.0056	123.88
UV/O3 PCL	4.3 ± 7.0	0.0124	55.91
Gelatin coated PCL	16.5 ± 24.1	0.0250	27.75

calculated every 12 h for 5 days. Cell growth kinetics for gelatin-coated PCL micro-carrier exhibited the highest final yield of 1.65×10^6 cell/mL (11-fold expansion) on the second day of cultivation. The culture showed fast growth (doubling time 27.75 h) before a lag phase at 36 h before the exponential phase (Fig. 4.1). The number of cells started to decrease 70 h later due to nutrient depletion and accumulation of waste in the culture media. Slow growth with a doubling time of 55.91 h was observed in UV/O$_3$ culture with UV/O$_3$-treated PCL. The maximum yield obtained was very low (4.25×10^5 cells/mL) compared to the gelatin-coated PCL microcarrier culture. This low yield could be attributed to the incompatibility of the surface as AFSC is a type of primary cell which may have low plating efficiency.

According to Hwang and coworkers (as cited in Amelia & Mohd Ridzuan, 2015, p. 18153), cells have different requirements for attachment depending on the cell's specificity. Therefore, incorporation of gelatin into PCL microcarrier surface may cater to the needs of AFSC as gelatin contains many reactive groups that allow binding to cell integrin leading to better attachment and growth. In contrast, UV/O$_3$-treated. PCL microcarrier offers only a charged surface. Besides, the immobilization of cells, particularly stem cells and primary cells, may enable the control of the behavior of the cells, including its differentiation potential. For example, Yang et al. (2003) showed that stem cells cultured on gelatin-coated microsphere maintain ex vivo expansion of stem cells while preserving their differentiation potential (Yang et al., 2003).

It is also important to establish a stable propagation of AFSC without differentiation during cultivation. According to Mun-Fun et al. (2015), AFSC is capable to differentiate into three types of germ layers (ectoderm, mesoderm, and endoderm). As such, spontaneous differentiation may give rise to a variety of phenotypes that may lead to the formation of teratomas (consisting of three tissue from the germ layers) at the transplantation site (Hentze et al., 2009). In a study by Murray and Edgar (2001), Leukemia Inhibitor Factor (LIF) was added to the culture medium to inhibit the differentiation of epiblast cell mouse embryonic stem cells into visceral and parietal (endoderm lineage). Therefore, it inhibits the formation of embryonic bodies during cultivation (Murray & Edgar, 2001).

Figure 4.2 shows the morphology of AFSC cells on untreated PCL, UV/O$_3$-treated PCL, and gelatin-coated PCL observed using phase contrast microscope and SEM at 60 h of cultivation. No aggregates were observed for all cultures for the first 3 days of cultivation, but cell bridges start to form in UV/O$_3$-treated PCL and gelatin-coated PCL microcarrier cultures.

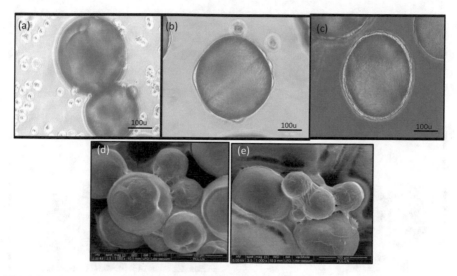

Fig. 4.2 Micrograph of AFSC at 96 h on (**a**) untreated PCL (**b**) UV/O$_3$ treated PCL (**c**) gelatin coated PCL visualized using an inverted phase contrast microscope (100 x amplification) Scale bar: 100 μm. SEM image of (**d**) AFSC on UV/O$_3$ treated PCL and (**e**) AFSC on gelatin-coated PCL (1000 x amplification) Scale bar: 100 μm

4.4.1 Spontaneous Differentiation of AFSC

The main goal of studying spontaneous differentiation of AFSC was to confirm that AFSC cells cultured for 5 days under stirred conditions using microcarriers retained their pluripotency. The functional pluripotency was determined by the ability of AFSC to differentiate spontaneously to form the good quality of embryonic bodies (EBs) (Itskovitz-Eldor et al., 2000). According to Mun-Fun et al. (2015), the formation of EBs is one of the principal tests to determine the "stemness" of the stem cell. In this phase of the study, a hanging drop method was applied in which the collected cells post trypsinization from UV/O$_3$-treated PCL and gelatin-coated PCL were resuspended in EB medium (without LIF), aliquoted to 20 μl drops containing 4000 cells, and placed on the underside of the petri dish lid. Trypsinized cells from untreated PCL microcarriers were not included in the formation of EB due to the insufficient number of cells to perform the spontaneous differentiation. Figure 4.3 illustrates the flow of AFSC cultivation on PCL microcarrier (UV/O$_3$-treated and gelatin-coated) and 2D culture (T-flask) as a positive control. Following is the spontaneous differentiation of the AFSC to the EB produced by the hanging drop method. Figures 4.3 (d and e) shows the unsuccessful spontaneous differentiation using different strains of AFSC into EB as a reference. Cells tend to be attached to the culture flask rather than to form cell aggregates (EBs).

EBs were analyzed based on smooth boundaries, size, and occurrence of the cavitation process (Kim et al., 2011). AFSC cultured on microcarriers (Figs. 4.3b and 4.3c) was able to form EBs with a size range between 100 and 300 μm. This

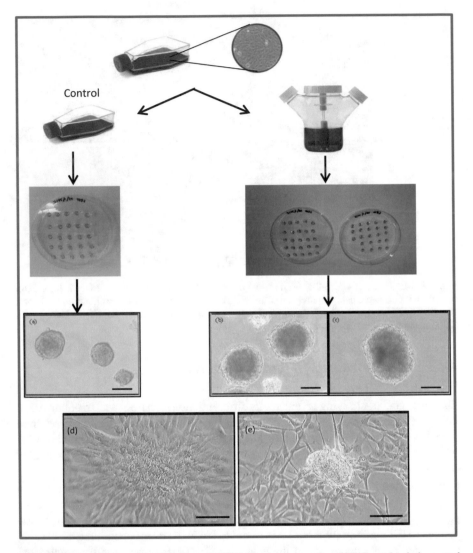

Fig. 4.3 Illustrative flow of cultivation of AFSC in spinner vessel on UV/O₃ and gelatin coated PCL microcarrier and T-flask culture as a control. The formation of EBs using hanging drop method from UV/O₃ (**b**) and gelatin coated (**c**) were compared to control (**a**). Figure (**d**) and (**e**) unsuccessful spontaneous differentiation to form EB. Scale bar: 100 μm

is a significant measure to ensure the quality and successful differentiation into the three cell lineages (Messana et al., 2008).

Besides, the formation of a cavity in the middle of EB is another measure of good EBs and its formation is significantly related to the initiation of EBs to differentiate into three germs layers (Rodda et al., 2002). According to Itskovitz-Eldor et al.

(2000), stem cells are true pluripotent because when they are capable to differentiate into all embryonic lineages. Thus, the formation of post-microcarrier EBs (3D) cultivation has confirmed that pluripotency of AFSC is being preserved; which is comparable to EBs of AFSC culture in 2D culture (T-flask).

4.5   Conclusion

This study explores the development of functional surface biodegradable PCL microcarrier and the use of PCL microcarrier in cell and tissue culture and regenerative medicine. The result shows that the gelatin-coated PCL microcarrier is capable to support the growth and proliferation of AFSC. The growth of AFSC was assisted only by the gelatin-coated PCL microcarrier as they are the primary type of cells with low plating efficiency that require supplementary growth factor to attach and proliferate in vitro. As a result of easy and safe accessibility, abundant cell numbers, and lack of ethical concerns, AFSC has emerged as an attractive source of stem cells for basic research and clinical applications. Compared to 2D cultures, AFSCs grown in 3D microenvironments of gelatin microcarrier had stable proliferation with a significantly higher expansion fold, suggesting that the gelatin-coated PCL microcarrier is an effective 3D support for anchorage-dependent AFSC.

Acknowledgements The authors are grateful to the Ministry of Higher Education Malaysia for their research grant (PRGS 11-001-0001) under the Prototype Research Grant Scheme (PRGS) and the Department of Biotechnology Engineering, International Islamic University Malaysia for their support.

References

Amelia, A. K., & Mohd Ridzuan, A. (2015). A review of cell adhesion studies for biomedical and biological applications. *International Journal of Molecular Sciences, 16,* 18149–18184.

Brun-Graeppi, A. K. A. S., Richard, C., Bessodes, M., Scherman, D., & Merten, D. W. (2011). Cell microcarrier and microcapsule of stimuli-responsive polymers. *Journal of Controlled Release, 149*(3), 209–224.

Butler, M. (2004). *The basic animal cell culture and technology.* Oxford University Press.

Chu, L., & Robinson, K. R. (2001). Industrial choice for protein production by large scale cell culture. *Biochemical Engineering, 12,* 180–187.

Freshney, R. I. (2010). *Culture of animal cells: A manual of basic technique* (6th ed.). Wiley-Blackwell.

Goh, T. K.-P., Zhang, Z.-Y., Chen, A. K.-L., Reuveny, S., Choolani, M., Chan, J. K. Y., & Oh, S. K.-W. (2013). Microcarrier culture for efficient expansion and osteogenic differentiation of human fetal mesenchymal stem cells. *BioResearch Open Access, 2*(2), 84–97. https://doi.org/10.1089/biores.2013.0001.

Hentze, H., Soong, P. L., Wang, S. T., Phillips, B. W., Putti, T. C., & Dunn, N. R. (2009). Teratoma formation by human embryonic stem cells: Evaluation of essential parameters for future safety studies. *Stem Cell Research, 2*(3), 198–210.

Itskovitz-Eldor, J., Schuldiner, M., Karsenti, D., Eden, A., Yanuka, O., Amit, M., Soreq, H., & Benvenisty, N. (2000). Differentiation of human embryonic stem cells into embryoid bodies compromising the three embryonic germ layers. *Molecular Medicine, 6*(2), 88–95.

Kim, J. M., Moon, S.-H., Lee, S. G., Cho, Y. J., Hong, K. S., Lee, J. H., & Chung, H.-M. (2011). Assessment of differentiation aspects by the morphological classification of embryoid bodies derived from human embryonic stem cells. *Stem Cells and Development, 20*(11), 1925–1935.

Liu, Q. (2004). Tissue engineering. In D. Shi (Ed.), *Biomaterials and tissue engineering.* Springer-Verlag Berlin Heidelberg.

Messana, J. M., Hwang, N. S., Coburn, J., Elisseeff, J. H., & Zhang, Z. (2008). Size of the embryoid body influences chondrogenesis of mouse embryonic stem cells. *Journal of Tissue Engineering and Regenerative Medicine, 2,* 499–506.

Mun-Fun, H., Ferdaos, N., Hamzah, S. N., Ridzuan, N., Hisham, N. A., Abdullah, S., & Nordin, N. (2015). Rat full term amniotic fluid harbors highly potent stem cells. *Research in Veterinary Science, 102,* 89–99.

Murray, P., & Edgar, D. (2001). The regulation of embryonic stem cell differentiation by leukaemia inhibitor factor (LIF). *Differentiation, 68*(4–5), 227–234.

Nikam, V. K., Gudsoorkar, V. R., Hiremath, S. N., Dolas, R. T., & Kashid, V. A. (2012). Microspheres-A novel grug delivery system: an overview. *International Journal of Pharmaceutical Chemical, 1*(1), 113–128.

Overstreet, M., Sohrabi, A., Polotsky, A., Hungerford, D., & Frondoza, C. G. (2003). Collagen microcarrier spinner culture promotes osteoblast proliferation and synthesis of matrix proteins. *Vitro Cellular and Developmental Biology, 39,* 228–234.

Rodda, S. J., Kavanagh, S. J., Rathjen, J., & Rathjen, P. D. (2002). Embryonic stem cell differentiation and the analysis of mammalian development. *International Journal of Developmental Biology, 46,* 449–458.

Sahil, K., Akanksha, M., Premjeet, S., Bilandi, A., & Kapoor, B. (2011). Microsphere: A review. *International Journal of Research in Pharmacy and Chemistry, 1*(4), 1184–1198.

Samsudin, N., Hashim, Y. Z. H-Y., Arifin, M. A., Mel, A., Salleh, H. M., Spoyan, I., & Jimat, D. N. (2017). Optimization of ultraviolet ozone treatment process for improvement of polycaprolactone (PCL) microcarrier performance. *Cytotech.* https://doi.org/10.1007/s10616-017-0071-x.

Seland, H., Gustafson, C.-J., Johnson, H., Junker, J. P. E., & Kratz, G. (2011). Transplantation of acellular dermis and keratinocytes cultured on porous biodegradable microcarriers into full-thickness skin injuries on athymic rats. *Burns: Journal of the International Society for Burn Injuries, 37*(1), 99–108.

Tan, H., Wu, J., Huang, D., & Gao, C. (2010). The design of biodegradable microcarriers for induced cell aggregation. *Macromolecular Bioscience, 10*(2), 156–163.

van der Velden-de Groot, C. a. (1995). Microcarrier technology, present status and perspective. *Cytotechnology, 18*(1–2), 51–6.

Wezel, V. (1967). Growth of cell-strains and primary cells on micro-carriers in homogeneous culture. *Nature, 216,* 64–65.

White, L. A., & Ades, E. W. (1990). Growth of vero E-6 cells on microcarriers in a cell bioreactor. *Journal of Clinical Microbiology, 28*(2), 283–286.

Yang, Y., Porte, M., Marmey, P., El Hai, A. J., Amedee, J., & Baquey, C. (2003). Covalent bonding of collagen on poly(L-lactic acid) by gamma irradiation. *Nuclear Instrumentation Methods, 207,* 165–174.

Chapter 5
Application of Spectroscopic and Chromatographic Methods for the Analysis of Non-halal Meats in Food Products

Abdul Rohman and Nurrulhidayah Ahmad Fadzillah

Abstract Meat is one of the major food groups which is frequently adulterated for economic reasons. In line with halal certification, the presence of non-halal meats in food products must be identified. Mislabelling and undeclaring meat types are typically seen in meat-based products such as meatballs, burgers, sausages and Salami. Therefore, the application of analytical methods for the detection, identification, and confirmation of non-halal meats is a must. This chapter highlighted the application of spectroscopic and chromatographic-based methods combined with several chemometrics for the identification and confirmation of non-halal meats in food products. Spectroscopy offered more accelerating methods with higher accuracy rates throughout the screening process for non-halal meats. Further confirmation can be done using chromatography by identifying specific markers present in analyzed non-halal meats.

Keywords Halal authentication · Pork · Dog meat · Chemometrics · Molecular spectroscopy

5.1 Introduction

Meat is considered one of the most commonly consumed foods worldwide due to the nutritional composition found in meats, especially protein. Meats contain essential healthy nutrients and are an excellent source of protein (Hassoun et al., 2020). From an Islamic perspective, meats could be classified as halal-meats that are allowed to

A. Rohman
Centre of Excellence, Institute for Halal Industry & Systems, Universitas Gadjah Mada, Yogyakarta 55281, Indonesia

Department of Pharmaceutical Chemistry, Faculty of Pharmacy, Universitas Gadjah Mada, Yogyakarta 55281, Indonesia

N. A. Fadzillah (✉)
International Institute for Halal Research and Training (INHART), International Islamic University Malaysia (IIUM), 53100 Gombak, Selangor, Malaysia
e-mail: nurrulhidayah@iium.edu.my

© The Author(s), under exclusive license to Springer Nature Switzerland AG 2021
A. Amid (ed.), *Multifaceted Protocols in Biotechnology, Volume 2*,
https://doi.org/10.1007/978-3-030-75579-9_5

be consumed and non-halal meats that are prohibited to be consumed according to Syariah law (Islamic jurisprudence) (Rohman & Windarsih, 2020).

With the increased awareness among the Muslim population to only consume halal and *tayyib (good or pure)* foods, the demand for halal food products also increased, emplacing a huge responsibility on the government, jurisprudence and manufacturers to assure the halalness of food products through halal certifications (Martuscelli et al., 2020). Indonesia has the regulation (Indonesian Act) No. 33 established in 2014 that mandates a halal certification for all halal products which are declared halal. Indeed, meat is one of most essential food components in the food industry and due to the big discrepancy between halal and non-halal meats, some unethical producers intentionally replace halal with non-halal meat (Nakyinsige et al., 2012).

At present, it is estimated that the Muslim population will increase from 1.8 billion in 2014 to 2.2 billion by 2030, with an approximate increased growth of 26.4%. The Muslim population occupies a quarter part of the world; therefore, it is not surprising if the Halal market exhibited a lucrative and significant impact on international business (Adiarni & Fortunella, 2018). However, there are several issues regarding Halal as non-halal meats either because of their source (such as pork) or from the way the animals were slaughtered. Disputes regarding the halal status of animal sources typically arise from thoughts' scholar (*madzhab*) to be followed by Muslim communities (Rohman et al., 2020), while debates on the halal status of various slaughtering methods arise due to the possibility that animals' slaughtering processes did not meet the Halal requirements determined by the Syariah law (Ali et al., 2020).

5.2   Meat's Authentication Using Spectroscopic and Chromatographic-Based Methods

In recent years, consumers have become more concerned about the quality, halalness, and safety of meat-based food products, especially the traceability and authenticity of meat sources (Hassoun et al., 2020). The analytical methods for meat-species identification and detection of adulteration are always needed for quality control and the safety of consumers. The Gold standard method used for halal meat authentication is a type of DNA-based methods, especially analytical methods applying polymerase chain reaction (PCR) with its development and revolution, including real-time PCR using primer species-specific, TaqMan probe and multiplex (Ali et al., 2014; Erwanto et al., 2018a, 2018b; Fajardo et al., 2010; Kumar et al., 2015; Rodríguez-Ramírez et al., 2011). However, PCR techniques involve more complex steps such as starting primer selection, denaturation of DNA, primer annealing and amplification. Therefore, a range of uncomplicated and quick methods were developed and used for analyzing non-halal meats in food products. Some reviews on the application of molecular spectroscopic (Esteki, Shahsavari, et al., 2018; Li et al., 2019; Rodriguez-Saona & Allendorf, 2011) and chromatographic methods (Abbas et al.,

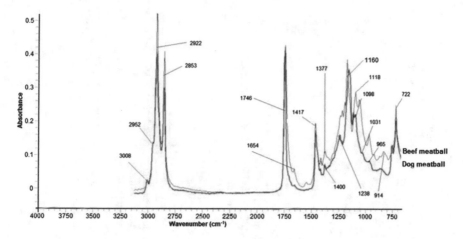

Fig. 5.1 Typical FTIR spectra of lipid components extracted from meats (beef and dog meat) in meatball samples

2018; Bosque-Sendra et al., 2012; Esteki, Simal-Gandara, et al., 2018) for food authentication and traceability have existed. In this chapter, two groups of methods (spectroscopy and chromatography) were critically assessed for the authentication analysis of non-halal meats in fresh and meat-based food products.

Among the molecular spectroscopic-based methods, Fourier transform infrared (FTIR) spectroscopy is the most applied technique for the analysis of non-halal meats either in fresh meats or meats in food-based products. The analysis of raw meats and meat-based foods is done by extracting the lipid fractions of meats using several extraction methods and then subjecting them to FTIR spectral measurement (Rohman, 2019). Typical FTIR spectra of lipid components extracted from meats are depicted in Fig. 5.1. The intensity is represented by absorbance values (not in transmittance mode) because absorbance can contain quantitative information according to Lambert–Beer law. Each peak and shoulder with specific wavenumbers in FTIR spectra represented functional groups present in lipids which can be correlated with compounds or a group of compounds composing the lipid. The main components of lipids extracted from meats which is represented by functional groups of methyl- (CH_3-) present in wavenumbers ($1/\lambda$) of 2953 cm^{-1} (str asymmetric), 2875 cm^{-1} (str symmetric), 1376 cm^{-1} (ben), methylene (-CH_2-) at $1/\lambda$ 2922 cm^{-1} (str asymmetric), 2853 cm^{-1} (str symmetric), 1462 cm^{-1} (ben), carbonyl- (C = O) at 1741 cm^{-1} (str) 1654 cm^{-1} corresponding to cis C = C str, 1417 cm^{-1} from = C–H cis disubstituted olefins str, and C–O (ether) at $1/\lambda$ 1117 and 1098 cm^{-1}, 965 cm^{-1} due to -HC = CH-(trans) in which str stands for stretching, while ben stands for bending vibration (Lerma-García et al., 2010).

The peaks and shoulders in FTIR spectra are rather complex, and they are used as variables for extracting the qualitative and quantitative information regarding the presence and the quantity of certain meats in food products. Therefore, data management using chemometrics was normally used (Nunes, 2014). Table 5.1 compiled the

Table 5.1 The application of different spectroscopic techniques used for the analysis of pork fresh meat or in meat-based food products

Non-halal meat	Methods	Issues	Chemometrics	Results	References
Pork	Visible and NIR spectroscopy using wavelength 400–2500 nm (vis) and 1/λ 25,000–4000 cm^{-1}	Analysis of pork in fresh and frozen-thawed minced beef	LDA and PLS-DA for discrimination and PLSR for quantitative analysis of pork	LDA and PLS-DA could discriminate minced beef and that adulterated with pork with 100% accuracy. PLSR could quantify pork with R^2 of 0.94–0.96 and SEP of 2.08–5.39% (w/w)	(Morsy & Sun, 2013)
Pork	On-site NIR spectrometer at 1/λ 6028–5480 cm^{-1}	Analysis of pork adulteration in sausages	PCA and SVM, spectra were preprocessed using 2nd derivative and SNV	PCA and SVM offered the correct classification between authentic and adulterated pork	(Schmutzler et al., 2015)
Pork	NIR spectra at a wavelength of 1000–2500 nm	Quantitative Analysis of pork as an adulterant in beef and chicken meatballs	PLSR for quantification	PLSR could facilitate the prediction of pork in beef meatballs with R^2 of 0.88, SECV of 3.45% and bias of 0.14%. For pork in chicken meatballs, R^2 of 0.83, SECV of 4.18%, and a bias of 0.22%, respectively	(Vichasilp & Poungchompu, 2014)
Pork	Vis/NIR spectra at 900–1700 nm	Quantitative analysis of pork in beef	PLSR for quantification	PLSR could successfully quantify pork in a binary mixture with beef with high R^2 and low RPD	(Rady & Adedeji, 2018)

(continued)

Table 5.1 (continued)

Non-halal meat	Methods	Issues	Chemometrics	Results	References
Pork	Visible-NIR hyperspectral imaging spectroscopy at wavelength 400–1000 nm	Analysis of pork lung adulteration in ground pork	PLSR for prediction of ground pork	PLSR using absorbances at 11 wavelengths based on loading plot (466, 480, 531, 562, 570, 617, 661, 729, 750, 816 and 982 nm) could predict the levels of ground pork with R^2 of 0.98, RMSEP of 4.47%, RPD of 7.16, and LOD of 7.58%	(Jiang et al., 2021)
Pork	NIR spectra at 700–850	Analysis of pork as adulterant in beef meatballs	LDA for classification and PLSR for quantification	LDA using the first derivative spectra could accurately classify beef meatballs and those adulterated with pork with an accuracy level of 100%. The prediction results of pork levels using NIR spectroscopy-PLS was in agreement with those using the immune-chromatographic method	(Kuswandi, Cendekiawan, et al., 2015)

(continued)

Table 5.1 (continued)

Non-halal meat	Methods	Issues	Chemometrics	Results	References
Pork	FTIR spectroscopy at mid-infrared region at 1/λ 4000–650 cm^{-1}	Analysis of pork as an adulterant in beef jerky	LDA, SIMCA, SVM	Comparing SIMCA and SVM, LDA offered the best model with an accuracy level of 100%. The LDA model was also in good agreement with the ELISA method	(Kuswandi, Putri, et al., 2015)
Pork	FTIR spectroscopy at 1/λ 4000–650 cm^{-1}	Rapid discrimination of pork in non-Halal Chinese ham sausages	PLS-DA and LS-SVM. Spectra were pretreated with SNV	PLS-DA and LS-SVV could classify halal and non-halal sausages containing pork. The sensitivity and specificity were 0.913 and 0.929 (PLS-DA) as well as 0.957 and 0.929 (LS-SVM)	(Xu et al., 2012)
Pork	FTIR spectra using 1/λ 1018–1284 cm^{-1} and 1200–1000 cm^{-1}	Analysis of pork in meatball broth	PCA (classification) and PLSR for quantification	PLSR using absorbances at 1018–1284 cm^{-1} could predict the fats extracted from pork meatballs. PCA using absorbances at 1200–1000 cm^{-1} could classify lipid components extracted from pork and beef meatballs	(Kurniawati et al., 2014)

(continued)

Table 5.1 (continued)

Non-halal meat	Methods	Issues	Chemometrics	Results	References
Pork	FTIR spectra at $1200-1000$ cm^{-1}	Analysis of beef meatballs adulterated with pork	PLSR for quantification	PLSR using absorbance values at $1200-1000$ cm^{-1} could predict pork contents with R^2 value of 0.997 and SEC of 0.04%	(Rohman et al., 2017)
Pork	FTIR spectroscopy	Quantitative analysis of pork in the mixture with camel and buffalo meats	PLSR for quantification	The relationship between the levels of pork in camel and buffalo meats and absorbance ratio ($R_{1464/1448}$ cm^{-1}) offered $R^2 > 0.99$ with acceptable errors	(Lamyaa, 2013)
Pork	FTIR spectroscopy	Identification of pork in the mixture with beef and mutton	PLS-DA and SVM	PLS-DA and SVM using FTIR spectra pretreated with normalization could discriminate beef/ mutton and that adulterated with pork with 100% accuracy	(Yang et al., 2018)
Pork	On-site NIR spectroscopy	Analysis of pork in a binary mixture of pork-beef, pork-chicken and ternary mixture of pork- beef-chicken	PLSR and SVR for quantification	Meats in binary and ternary mixtures could be quantified using PLSR and SVR with acceptable recovery (high R^2) and precision (low errors)	(Silva et al., 2020)
Wild boar meat (WBM)	FTIR spectroscopy at $1250-1000$ cm^{-1}	Analysis of WBM in beef meatballs	PCA for classification and PLSR for quantification	PCA could classify beef meatballs and those adulterated with WBM. The levels of WBM could be predicted with acceptable accuracy and precision	(Guntarti et al., 2015)

(continued)

Table 5.1 (continued)

Non-halal meat	Methods	Issues	Chemometrics	Results	References
WBM	FTIR spectroscopy at $1250-1000\ cm^{-1}$	Analysis of WBM in beef meatballs	PCA for classification and PLSR for quantification	The equation obtained equaled the predicted value = $0.994 \times$ actual value + 0.334; with R^2 of 0.998, RMSEC of 1.22% and RMSECV of 2.68%. PCA could classify WBM and beef meatballs	(Sari & Guntarti, 2018)

$1/\lambda$ = wavenumbers; LDA = Linear Discriminant Analysis; SIMCA = soft independent modelling of class analogy; SVM = Support Vector Machines; LS-SVM = least squares support vector machine; SEC = standard error of calibration; SEP = standard error of prediction; RPD = relative percentage difference

application of different spectroscopic techniques used for the analysis of pork fresh meat or in meat-based food products.

5.3 Analysis of Non-halal Meats in Fresh and Meat-Based Food Products

There are some non-halal meats that are not allowed to be consumed by Muslim communities, namely, pork (pig meat), wild boar meat (WBM), dog meat (DM), monkey meat, cat meat, snake meat, as well as meats from some amphibians and other animals which live in the water and on the ground such as frogs and crocodiles (Ceranic & Bozinovic, 2009).

5.3.1 Analysis of Pork in Fresh and Meat-Based Food

Pork is the most used non-halal meat in food products, and the analytical methods intended for detecting pork in raw components in the mixture with other meats and in meat-based food products such as hamburger, meatballs, and sausages are widely reported. Near-infrared (NIR), mid-infrared (MIR) spectroscopy, gas chromatography and liquid chromatography in combination with some chemometrics techniques have been validated and applied in the analysis of pork for halal authentication. MIR spectra combined with PLS-Kernel have been successfully used for quantifying pork in beef with a limit of detection reaching up to 1.4% wt/wt. The chemometrics of the PLS-Kernel algorithm exhibited good performance results for treating many variables in complex spectral data due to the lower memory storage needed with fewer computation cycles. The absorbance values at wavenumbers of 1900–900 cm^{-1} were used as variables while preparing the calibration curve for the relationship between actual values and MIR-PLS Kernel predicted values, resulting in the statistical performance of R^2 of 0.9994 and a relative error of prediction (REP) of 3.7%. High R^2 and low REP indicated that MIR spectroscopy combined with PLS-Kernel provided an accurate and precise technique for quantifying pork (non-halal meat) in beef (halal meat) for halal authentication purposes (Abu-Ghoush et al., 2017).

Fourier transform near-infrared spectroscopy (FT-NIR) used in combination with chemometrics of PCA, Partial Least-Squares Discriminant Analysis (PLS-DA) and PLSR has been currently used for the analysis of pork in other meats (lamb, chicken, mutton, beef, camel). PCA was utilized for exhibiting the similarities and differences of pork and other types of meat. PLS-DA was used for the discrimination of pork and other types of meats, while the multivariate calibration of PLSR was applied for the prediction of pork levels in the mixture with other meats. This study involved a large number of samples (>5900 samples from different countries) scanned at wavenumbers of 10,000–4000 cm^{-1}. PCA using full NIR spectra that resulted in PC-1 and PC-2 explaining 87 and 8% of the total spectral variability. From PC1 and PC2 curves, it is clear that PCA revealed a complete segregation between pork and other types of meat. All pork samples were grouped in two different categories corresponding to fresh pork and treated pork (dried and smoked pork samples). PLS-DA could distinguish pork from other meats successfully where the lowest level of pork that could be detected is 10%. The prediction of pork levels in the mixture with other meats was optimized using full NIR spectra preprocessed with standard normal variate (SNV) and unit vector normalization transformation. This better model showed a good correlation between actual values and FT-NIR predicted values with R^2 of 0.9774, RMSECV of 1.08% and RMSEP of 1.835%. From these results, it is evident that the combination of NIR spectra and chemometrics of PCA, PLS-DA and PLSR is an effective mechanism for the authentication analysis of distinguishing meats from non-halal meat containing pork (Mabood et al., 2020).

Infrared spectroscopy in mid-region (MIR) combined with chemometrics of PCA, SIMCA and PLS has statistically reported that fats extracted from pork (pig meat) are used as an adulterant in pure ghee (heat clarified milk fat) either qualitatively or quantitatively. Optimizing wavenumbers is capable of providing a good classification using PCA and SIMCA. The combined wavenumbers region of 3030–2785, 1786–1680, and 1490–919 cm^{-1} which exhibit strong intensities (absorbances) were utilized for these analytical tasks. SIMCA, using this combined region could classify lard, pure ghee and lard mixed with ghee with efficiency of 100%. PCA, using PC1 accounting of 92% and PC2 accounting of 8% variances applying absorbance values at 3030–2785 cm^{-1} as variables, was capable of clustering pure pork, pure ghee (cow and buffalo), and pure ghee adulterated with pork fat (lard) with different concentrations as shown in Fig. 5.2. The presence of lard of about 5% in pure ghee revealed a clear separation, while lard with a lower rate of 5% (especially 3%) displayed an overlapping separation. The wavenumbers used for PCA were also successfully applied for the prediction of lard in pure ghee with R2 for the correlation between the actual and FTIR predicted value of > 0.99 and a detection limit of 3% (Upadhyay et al., 2018).

Raman spectroscopy combined with PCA is also a successful approach for a swift identification of pork meat and other meats (sheep, cattle, fish, goat, poultry, and buffalo) and in salami products in a shorter period of time (about 30 s) after the extraction of fats from the corresponding meats. PCA was performed using variables of absorbance values at 200–2000 cm^{-1}. The first two PCs (PC1 and PC2) having the highest variances with 65.86 and 20.98% were chosen, and the plotting of PC1 and PC2 was capable of identifying pork and other meats in Salami products (Boyaci et al., 2014).

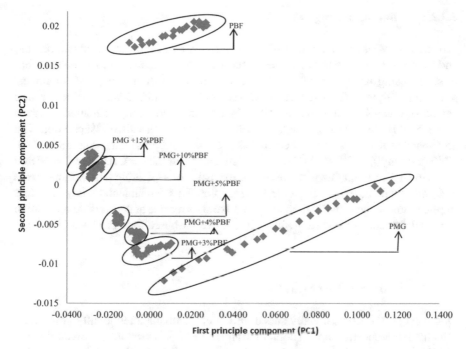

Fig. 5.2 PCA based on principal component (PC1) and second principal component (PC2) using absorbance values of 3030–2785 cm^{-1} for clustering pure pork (PBF), pure mutton ghee (PMG), and PMG adulterated with PBF with different concentrations. Adopted from (Upadhyay et al., 2018)

5.3.2   Analysis of Wild Boar Meat as Meat Adulterants

Currently, Wild boar meat, which is similar to pork, has been reported to be used as an adulterant in beef meatballs in Indonesia and Malaysia. Therefore, FTIR spectroscopy is developed for the identification and quantification of WBM in meatballs. The selection of fingerprint regions in FTIR spectra was typically carried out before modeling using chemometrics techniques. Some studies usually involve the addition of WBM into beef meatballs with varying concentrations. PCA using PC1 and PC2 based on absorbance values at combined wavenumbers of 999–1481 and 1650–1793 cm^{-1} has successfully distinguished WBM and beef meatballs. In addition, PLSR using the first derivative spectra at the combined wavenumbers of 999–1481 and 1650–1793 cm^{-1} could predict the levels of WBM with R^2 value of 0.9991, RMSEC of 1.028%, and RMSECV of 0.30%, respectively. This result indicated that FTIR spectroscopy combined with PCA and PLSR is effective for authenticating halal meatballs from WBM (Ahda et al., 2020).

5.3.2.1 Analysis of Frog Meats

Frog meat (FM) is a popular food also known as "*swike*" (in Indonesia and Malaysia) and is considered a non-halal meat according to *madzhab Syafi'i*. FTIR spectroscopy using unsupervised pattern recognition of PCA was applied for fingerprinting profiling of fats extracted from FM and other animal fats, marine oils and vegetable oils. PCA using PC1 and PC2, which are accounting for 88 and 7% variances, could classify fats extracted from FM and others with a distinct separation. The absorbance values at wavenumbers of 2922 (corresponding to -CH$_2$-), 2853 (CH2-), 1745 (carbonyl), 1158 cm^{-1} (C–O), and 721 cm^{-1} (*cis*–CH = CH) were considered as the most discriminating variables that are capable of separating fats extracted from FM and others, as explained in the loading plot. This fundamental study has clear implications on the identification of FM fat from its marine and vegetable oils for the potential detection of FM adulteration in various fat-based food products (Ali et al., 2015).

5.3.2.2 Analysis of Dog Meat

Dog meat (DM) is occasionally added into beef meatballs intentionally or unintentionally. FTIR spectroscopy in combination with PLSR is used for the quantification of DM in beef meatballs. During the preparation of the prediction and validation samples, DM was added into beef meatballs in the range of 0–100% wt/wt. All samples were subjected to fat extraction using the Folch method applying chloroform–methanol (2:1 v/v) as extracting solvents, and the lipid extracts obtained were scanned at 4000–650 cm^{-1}. PLSR applying absorbance values at combined wavenumbers region of 1782–1623 and 1485–659 cm^{-1} using normal spectra with a high accuracy as indicated by high R^2 either in the calibration model (0.993) or in the validation model (0.995). The errors were also low in both calibration (RMSEC of 1.63%) and in cross validation using the leave-one-out technique with a RMSECV of 2.68%. FTIR spectroscopy combined with PLSR provided an accurate and precise technique for the quantitative analysis of DM in beef meatballs (Rahayu, Rohman, et al., 2018).

The classification and quantification of DM in beef sausages have been carried out. Based on the optimization in terms of FTIR spectra model and wavenumbers region, the authors discovered that the normal spectra at wavenumbers of 1124–688 cm^{-1} was suitable for the quantification of DM, which produced the equation of predicted value = 0.9999 × actual value + 0.0004, with statistical parameters of R^2 of 0.9999, RMSEC of 0.30%, RMSEP of 0.05% and RMSECV of 0.05%. Based on PCA results, samples 1 and 5 (sausages obtained from a commercial market) had the closest point to that of DM (Guntarti & Ayu Purbowati, 2019).

Rahayu et al. have compared two extraction techniques, namely Bligh-Dyer and Folch methods for fat extraction from dog meat in meatballs formulation. The lipids obtained were scanned in FTIR spectrophotometer at the mid-IR region (4000–650 cm^{-1}). FTIR spectral bands correlated with fats of beef, dog and the mixture of

DM and beef were analyzed. The small variations among spectra were utilized as basic tools to differentiate between DM and beef. PCA using variables of absorbance values at 1700–700 cm^{-1} was capable of identifying DM in meatballs (Rahayu, Martono, et al., 2018).

5.3.2.3 Analysis of Rat Meat

Our group has reported the application of FTIR spectroscopy combined with chemometrics of PCA and PLSR for qualitative (identification) and quantitative analysis of rat meat (RM) in beef meatballs. Lipids in meatballs were extracted using Soxhlet, applying hexane as an extracting solvent according to the AOAC method. Lipids components scanned at mid-IR region (4000–650 cm^{-1}) were classified using PCA for grouping the meatballs composed of beef and rat meat. In this study, some wavenumber regions were optimized for getting to acquire the best classification and prediction models. Finally, absorbances at wavenumbers of 750–1000 cm^{-1} were preferred for PCA PLSR. PCA using PC1 and PC2 could classify beef and rat meatballs. In addition, for the quantitative analysis, the R^2 value showing the correlation between the actual values and FTIR predicted values was 0.993 with the obtained equation for such relationship of predicted value $= 0.9417 \times$ actual value $+ 2.8410$ with RMSEC of 1.79% (Rahmania et al., 2015).

Assisted with chemometrics of PCA and multivariate calibrations, FTIR spectroscopy is reported to have been used for the identification and quantification of rat meat (RM) in food products such as sausages. Some lipid extraction techniques were carried out during this task, including Bligh and Dyer, Folch, and Soxhlet. The lipid extracted was subjected to FTIR measurement at the wavenumbers region of 4000–650 cm^{-1}. PCA was successfully applied for the classification of fats extracted from RM sausages and fats extracted with different extraction methods from beef sausages using variables of FTIR spectral absorbances at 1/λ 1800–750 cm^{-1} with PC1 and PC2 contribution and Soxhlet was 97.57% and 1.28% (Bligh and Dyer), 85.50 and 10.64% (Folch) and 97.86 and 2.02% (Soxhlet). In addition, quantitative analysis of RM in sausages extracted using three extraction techniques revealed good accuracy results as indicated by a high coefficient of determination (R^2) between actual values of RM prepared in sausages and FTIR predicted values, namely using the Soxhlet method. The results were 0.945 (Bligh and Dyer), 0.991 (Folch), and 0.992 (Soxhlet). The precision method was also acceptable as indicated by the low values of error with RMSEC of 2.73, 1.73 and 1.69%, respectively, using Bligh and Dyer, Folch, and Soxhlet. In validation models using sausage samples extracted by Folch and Soxhlet, the R^2 values and RMSEP values obtained were 0.458 and 18.90% (Folch), and 0.983 and 4.21 (Soxhlet). Guntarty et al. also used a similar method (FTIR spectra combined with PCA and PLSR) for the identification and quantification of RM in meatball samples using absorbance values at wavenumbers of 1800–750 cm^{-1}. The extraction method used was Soxhlet using hexane as an extracting solvent (Guntarti & Prativi, 2017). Meatballs from RM and beef could also be differentiated using FTIR spectroscopy combined with two unsupervised pattern recognitions of PCA and cluster analysis employing variables of absorbance values at 4000–400 cm^{-1} (Rosyidi & Khamidinal, 2019).

The analysis of fatty acid composition using gas chromatography with a flame ionization detector exhibited significant differences between the contents of fatty acids in fats extracted from beef and rat meat sausages. The contents of fatty acids of C12:0, C16:0, C16:1 cis 9, and C18:0 in rat lipid were higher than those in beef lipid, and the contrary was observed for fatty acids of C14:1 cis 9, C15:0, C17:0, C17:1 cis 10, unsaturated C18, and C21:0. The difference in fatty acid compositions can be used as complementary data for determining the presence of RM in food products (Pebriana et al., 2017). Fatty acid compositions determination by gas chromatography with a mass spectrometric detector combined with PCA's chemometrics were successfully applied for the differentiation of fats extracted from RM and other fats from pigs, cows, chickens, wild boars, dogs, and goats. The dominant fatty acids extracted from Wistar RM were oleate (40.48%), followed by linoleate (30.14%), palmitate (19.08%), stearate (2.55%), palmitoleate (0.73%) and myristate (0.1509%) (Guntarti et al., 2020). A similar procedure was also applied for the analysis of fats extracted from RM strain Sprague Dawley (Guntarti et al., 2021) and black rats (Utami et al., 2018). The combination of FTIR spectroscopy and fatty acid composition could be an effective method for detecting the adulteration of rat meat into beef meatballs (Damayanti et al., 2020).

5.3.2.4   Analysis of Donkey Meat

In China, the adulteration of donkey meat products with a similar species, namely horse meat is becoming a widespread concern. Therefore, the availability of analytical methods for detecting donkey meat is a must (Wang et al., 2020). The analysis of donkey meat as an adulterant in beef meatballs was carried out using FTIR spectroscopy combined with chemometrics of hierarchical cluster analysis (HCA). Before being used for clustering, FTIR spectra were subjected to some spectral treatments. Finally, FTIR spectra were preprocessed using first derivatization, first derivatization + vector normalization and vector normalization at wavenumbers region of 1480–1425 cm^{-1} which were capable of distinguishing donkey-adulterated beef meatballs, donkey meatballs and beef meatballs with 100% sensitivity and specificity (Candoğan et al., 2020).

The combination of near-infrared (NIR) spectroscopy and chemometrics has been used for differentiating donkey meat (167 samples) from different parts of the donkey's body, beef, pork and mutton with reflectance models at wavenumbers of 12,500–4000 cm^{-1}. The chemometrics of soft independent modeling of class analogy (SIMCA) and least squares-support vector machine (LS-SVM) were applied. Some NIR spectral treatments, including Savitzky-Golay smoothing and derivative spectra (first and second-derivative spectra), multiplicative scatter correction and standard normal variate, were compared and used. LS-SVM using the first 6PCs and SIMCA with the first 8PCs could accurately classify donkey meat and other meats with an accuracy of 100% in the calibration set and 98.96% in prediction sets (LS-SVM) and

100% in calibration and 97.53% in prediction sets (SIMCA). These results indicated that NIR spectra combined with chemometrics methods offered fast and reliable tools for distinguishing DM from other meats (Xiao-ying et al., 2014).

References

Abbas, O., Zadravec, M., Baeten, V., Mikuš, T., Lešić, T., Vulić, A., Prpić, J., Jemeršić, L., & Pleadin, J. (2018). Analytical methods used for the authentication of food of animal origin. *Food Chemistry, 246*(October 2017), 6–17. https://doi.org/10.1016/j.foodchem.2017.11.007.

Abu-Ghoush, M., Fasfous, I., Al-Degs, Y., Al-Holy, M., Issa, A. A., Al-Reyahi, A. Y., & Alshathri, A. A. (2017). Application of mid-infrared spectroscopy and PLS-Kernel calibration for quick detection of pork in higher value meat mixes. *Journal of Food Measurement and Characterization, 11*(1), 337–346. https://doi.org/10.1007/s11694-016-9402-4.

Adiarni, N., & Fortunella, A. (2018). *The analysis of halal assurance system implementation (HAS 23000) in fried chicken flour product: A case study on XXX brand.* In (*Icosat 2017*) (pp. 57–61). https://doi.org/10.2991/icosat-17.2018.13.

Ahda, M., Guntarti, A., Kusbandari, A., & Melianto, Y. (2020). Authenticity analysis of beef meatball adulteration with wild boar using FTIR spectroscopy combined with chemometrics. *Journal of Microbiology, Biotechnology and Food Sciences, 9*(5), 937–940. https://doi.org/10.15414/jmbfs.2020.9.5.937-940.

Ali, M. E., Nina Naquiah, A. N., Mustafa, S., & Hamid, S. B. A. (2015). Differentiation of frog fats from vegetable and marine oils by Fourier Transform Infrared Spectroscopy and chemometric analysis. *Croatian Journal of Food Science and Technology, 7*(1), 1–8. https://doi.org/10.17508/cjfst.2015.7.1.03.

Ali, M. E., Razzak, M. A., & Hamid, S. B. A. (2014). Multiplex PCR in species authentication: Probability and prospects—A review. *Food Analytical Methods, 7*(10), 1933–1949. https://doi.org/10.1007/s12161-014-9844-4.

Ali, N. S. M., Zabidi, A. R., Manap, M. N. A., Zahari, S. M. S. N. S., & Yahaya, N. (2020). Identification of metabolite profile in halal and non-halal broiler chickens using Fourier-transform infrared spectroscopy (FTIR) and ultra high performance liquid chromatography-time of flight-mass spectrometry (UHPLC-TOF-MS). *Malaysian Applied Biology, 49*(3), 87–93.

Bosque-Sendra, J. M., Cuadros-Rodríguez, L., Ruiz-Samblás, C., & de la Mata, A. P. (2012). Combining chromatography and chemometrics for the characterization and authentication of fats and oils from triacylglycerol compositional data-A review. *Analytica Chimica Acta, 724*, 1–11. https://doi.org/10.1016/j.aca.2012.02.041.

Boyaci, I. H., Uysal, R. S., Temiz, T., Shendi, E. G., Yadegari, R. J., Rishkan, M. M., Velioglu, H. M., Tamer, U., Ozay, D. S., & Vural, H. (2014). A rapid method for determination of the origin of meat and meat products based on the extracted fat spectra by using of Raman spectroscopy and chemometric method. *European Food Research and Technology, 238*(5), 845–852. https://doi.org/10.1007/s00217-014-2168-1.

Candoğan, P. D. K., Deniz, E., Güneş Altuntaş, E., Iğci, N., & Özel Demiralp, D. (2020). SiğirEti Karişimlarinda Domuz, At Ve EşekEtTağşişinin Fourier DönüşümlüKizilötesiSpektroskopisiİlBelirlenmesi. *Gida/The Journal of Food, 45*, 369–379. https://doi.org/10.15237/gida.gd19146.

Ceranic, S., & Bozinovic, N. (2009). Possibilities and significance of has implementation (Halal assurance system) in existing quality system in food industry. *Biotechnology in Animal Husbandry, 25*(3–4), 261–266. https://doi.org/10.2298/bah0904261c.

Damayanti, A., Djalil, A. D., Susanti, A. T., Apriani, B., & Astuti, I. Y. (2020). Analysis of rat adulteration in beef meatball using Fourier transform infrared spectroscopy and gas chromatography-mass spectrometry for halal authentication. In *IRCPAS/2020/OP-219, 8916* (Ircpas).

Erwanto, Y., Rohman, A., Arsyanti, L., & Pranoto, Y. (2018a). Identification of pig DNA in food products using polymerase chain reaction (PCR) for halal authentication—A review. *International Food Research Journal, 25*(4), 1322–1331.

Erwanto, Y., Rohman, A., Arsyanti, L., & Pranoto, Y. (2018b). Use of polymerase chain reaction to test for presence of pig derivatives in halal authentication studies. *International Food Research Journal, 25*(August), 1322–1331.

Esteki, M., Shahsavari, Z., & Simal-Gandara, J. (2018). Use of spectroscopic methods in combination with linear discriminant analysis for authentication of food products. *Food Control, 91,* 100–112. https://doi.org/10.1016/j.foodcont.2018.03.031.

Esteki, M., Simal-Gandara, J., Shahsavari, Z., Zandbaaf, S., Dashtaki, E., & Vander Heyden, Y. (2018, April). A review on the application of chromatographic methods, coupled to chemometrics, for food authentication. *Food Control, 93,* 165–182. https://doi.org/10.1016/j.foodcont.2018.06.015.

Fajardo, V., González Isabel, I., Rojas, M., García, T., & Martín, R. (2010). A review of current PCR-based methodologies for the authentication of meats from game animal species. *Trends in Food Science and Technology, 21*(8), 408–421. https://doi.org/10.1016/j.tifs.2010.06.002.

Guntarti, A., & Ayu Purbowati, Z. (2019). Analysis of dog fat in beef sausage using FTIR (Fourier Transform Infrared) combined with chemometrics. *Pharmaciana, 9*(1), 21–28. https://doi.org/10.12928/pharmaciana.v%vi%i.10467.

Guntarti, A., Gandjar, I. G., & Jannah, N. M. (2020). Authentication of wistar rat fats with gas chromatography mass spectometry combined by chemometrics. *Potravinarstvo Slovak Journal of Food Sciences, 14*(December 2019), 52–57. https://doi.org/10.5219/122910.5219/1229.

Guntarti, A., Martono, S., Yuswanto, A., & Rohman, A. (2015). FTIR spectroscopy in combination with chemometrics for analysis of wild boar meat in meatball formulation. *Asian Journal of Biochemistry, 10*(4). https://doi.org/10.3923/ajb.2015.165.172.

Guntarti, A., Ningrum, K. P., Gandjar, I. G., & Salamah, N. (2021). Authentication of sprague dawley rats (Rattus norvegicus) fat witH GC-MS (Gas chromatography-mass spectrometry) combined with chemometrics. *International Journal of Applied Pharmaceutics, 13*(2), 1–6. https://innova reacademics.in/journals/index.php/ijap/article/view/40130.

Guntarti, A., & Prativi, S. R. (2017). Application method of Fourier Transform Infrared (FTIR) combined with chemometrics for analysis of rat meat (Rattus Diardi) in meatballs beef. *Pharmaciana, 7*(2), 133. https://doi.org/10.12928/pharmaciana.v7i2.4247.

Hassoun, A., Måge, I., Schmidt, W. F., Temiz, H. T., Li, L., Kim, H. Y., Nilsen, H., Biancolillo, A., Aït-Kaddour, A., Sikorski, M., Sikorska, E., Grassi, S., & Cozzolino, D. (2020). Fraud in animal origin food products: Advances in emerging spectroscopic detection methods over the past five years. *Foods, 9*(8). https://doi.org/10.3390/foods9081069.

Jiang, H., Ru, Y., Chen, Q., Wang, J., & Xu, L. (2021). Near-infrared hyperspectral imaging for detection and visualization of offal adulteration in ground pork. *Spectrochimica Acta—Part A: Molecular and Biomolecular Spectroscopy, 249,* 1–9. https://doi.org/10.1016/j.saa.2020.119307.

Kumar, A., Kumar, R. R., Sharma, B. D., Gokulakrishnan, P., Mendiratta, S. K., & Sharma, D. (2015). Identification of species origin of meat and meat products on the DNA basis: A review. *Critical Reviews in Food Science and Nutrition, 55*(10), 1340–1351. https://doi.org/10.1080/104 08398.2012.693978.

Kurniawati, E., Rohman, A., & Triyana, K. (2014). Analysis of lard in meatball broth using Fourier transform infrared spectroscopy and chemometrics. *Meat Science, 96*(1), 94–98. https://doi.org/10.1016/j.meatsci.2013.07.003.

Kuswandi, B., Cendekiawan, K. A., Kristiningrum, N., & Ahmad, M. (2015). Pork adulteration in commercial meatballs determined by chemometric analysis of NIR Spectra. *Journal of Food Measurement and Characterization, 9*(3), 313–323. https://doi.org/10.1007/s11694-015-9238-3.

Kuswandi, B., Putri, F. K., Gani, A. A., & Ahmad, M. (2015). Application of class-modelling techniques to infrared spectra for analysis of pork adulteration in beef jerkys. *Journal of Food Science and Technology, 52*(12), 7655–7668. https://doi.org/10.1007/s13197-015-1882-4.

Lamyaa, M. A. (2013). Discrimination of pork content in mixtures with raw minced camel and buffalo meat using FTIR spectroscopic technique. *International Food Research Journal, 20*(3), 1389–1394.

Lerma-García, M. J., Ramis-Ramos, G., Herrero-Martínez, J. M., & Simó-Alfonso, E. F. (2010). Authentication of extra virgin olive oils by Fourier-transform infrared spectroscopy. *Food Chemistry, 118*(1), 78–83. https://doi.org/10.1016/j.foodchem.2009.04.092.

Li, Q., Chen, J., Huyan, Z., Kou, Y., Xu, L., Yu, X., & Gao, J. M. (2019). Application of Fourier transform infrared spectroscopy for the quality and safety analysis of fats and oils: A review. In *Critical reviews in food science and nutrition* (Vol. 59, Issue 22, pp. 3597–3611). Taylor and Francis Inc. https://doi.org/10.1080/10408398.2018.1500441.

Mabood, F., Boqué, R., Alkindi, A. Y., Al-Harrasi, A., Al Amri, I. S., Boukra, S., Jabeen, F., Hussain, J., Abbas, G., Naureen, Z., Haq, Q. M. I., Shah, H. H., Khan, A., Khalaf, S. K., & Kadim, I. (2020). Fast detection and quantification of pork meat in other meats by reflectance FT-NIR spectroscopy and multivariate analysis. *Meat Science, 163*(September 2019), 108084. https://doi.org/10.1016/j.meatsci.2020.108084.

Martuscelli, M., Serio, A., Capezio, O., & Mastrocola, D. (2020). Meat products, with particular emphasis on salami: A review. *Foods, 9*, 1–19.

Morsy, N., & Sun, D. W. (2013). Robust linear and non-linear models of NIR spectroscopy for detection and quantification of adulterants in fresh and frozen-thawed minced beef. *Meat Science, 93*(2), 292–302. https://doi.org/10.1016/j.meatsci.2012.09.005.

Nakyinsige, K., Man, Y. B. C., & Sazili, A. Q. (2012). Halal authenticity issues in meat and meat products. *Meat Science, 91*(3), 207–214. https://doi.org/10.1016/j.meatsci.2012.02.015.

Nunes, C. A. (2014). Vibrational spectroscopy and chemometrics to assess authenticity, adulteration and intrinsic quality parameters of edible oils and fats. *Food Research International, 60*, 255–261. https://doi.org/10.1016/j.foodres.2013.08.041.

Pebriana, R. B., Rohman, A., Lukitaningsih, E., & Sudjadi. (2017). Development of FTIR spectroscopy in combination with chemometrics for analysis of rat meat in beef sausage employing three lipid extraction systems. *International Journal of Food Properties, 20*(Supplement 1), 1995–2005. https://doi.org/10.1080/10942912.2017.1361969.

Rady, A., & Adedeji, A. (2018). Assessing different processed meats for adulterants using visible-near- infrared spectroscopy. *Meat Science, 136*(October 2017), 59–67. https://doi.org/10.1016/j.meatsci.2017.10.014.

Rahayu, S. W., Martono, S., Sudjadi, S., & Rohman, A. (2018). The potential use of infrared spectroscopy and multivariate analysis for differentiation of beef meatball from dog meat for Halal authentication analysis. *Journal of Advanced Veterinary and Animal Research, 5*(3), 307–314. https://doi.org/10.5650/jos.ess15294.

Rahayu, W. S., Rohman, A., Martono, S., & Sudjadi. (2018). Application of FTIR spectroscopy and chemometrics for halal authentication of beef meatball adulterated with dog meat. *Indonesian Journal of Chemistry, 18*(2), 376–381. https://doi.org/10.22146/ijc.27159.

Rahmania, H., Sudjadi, & Rohman, A. (2015). The employment of FTIR spectroscopy in combination with chemometrics for analysis of rat meat in meatball formulation. *Meat Science, 100*, 301–305.

Rodríguez-Ramírez, R., González-Córdova, A. F., & Vallejo-Cordoba, B. (2011). Review: Authentication and traceability of foods from animal origin by polymerase chain reaction-based capillary electrophoresis. *Analytica Chimica Acta, 685*(2), 120–126. https://doi.org/10.1016/j.aca.2010.11.021.

Rodriguez-Saona, L. E., & Allendorf, M. E. (2011). Use of FTIR for rapid authentication and detection of adulteration of food. *Annual Review of Food Science and Technology, 2*(1), 467–483. https://doi.org/10.1146/annurev-food-022510-133750.

Rohman, A. (2019). The employment of Fourier transform infrared spectroscopy coupled with chemometrics techniques for traceability and authentication of meat and meat products. *Journal of Advanced Veterinary and Animal Research, 6*(1). https://doi.org/10.5455/javar.2019.f306.

Rohman, A., Himawati, A., Triyana, K., Sismindari, & Fatimah, S. (2017). Identification of pork in beef meatballs using Fourier transform infrared spectrophotometry and real-time polymerase chain reaction. *International Journal of Food Properties, 20*(3), 654–661. https://doi.org/10.1080/10942912.2016.1174940.

Rohman, A., & Windarsih, A. (2020). The application of molecular spectroscopy in combination with chemometrics for halal authentication analysis: A review. *International Journal of Molecular Sciences, 21*(14), 1–18. https://doi.org/10.3390/ijms21145155.

Rohman, A., Windarsih, A., Erwanto, Y., & Zakaria, Z. (2020). Review on analytical methods for analysis of porcine gelatine in food and pharmaceutical products for halal authentication. *Trends in Food Science and Technology, 101*(May), 122–132. https://doi.org/10.1016/j.tifs.2020.05.008.

Rosyidi, N. N., & Khamidinal. (2019). Analisis Lemak Bakso Tikus dalam Bakso Sapi di Sleman Menggunakan Spektroskopi Inframerah (Fourier Transform Infrared). *Indonesian Journal of Halal Science, 001*(01), 18–29.

Sari, T. N. I., & Guntarti, A. (2018). Wild boar fat analysis in beef sausage using Fourier Transform Infrared method (FTIR) combined with chemometrics. *Jurnal Kedokteran Dan Kesehatan Indonesia, 9*(4), 16–23.

Schmutzler, M., Beganovic, A., Böhler, G., & Huck, C. W. (2015). Methods for detection of pork adulteration in veal product based on FT-NIR spectroscopy for laboratory, industrial and on-site analysis. *Food Control, 57*, 258–267. https://doi.org/10.1016/j.foodcont.2015.04.019.

Silva, L. C. R., Folli, G. S., Santos, L. P., Barros, I. H. A. S., Oliveira, B. G., Borghi, F. T., dos Santos, F. D., Filgueiras, P. R., & Romão, W. (2020, April). Quantification of beef, pork, and chicken in ground meat using a portable NIR spectrometer. *Vibrational Spectroscopy, 111*. https://doi.org/10.1016/j.vibspec.2020.103158.

Upadhyay, N., Jaiswal, P., & Jha, S. N. (2018). Application of attenuated total reflectance Fourier Transform Infrared spectroscopy (ATR–FTIR) in MIR range coupled with chemometrics for detection of pig body fat in pure ghee (heat clarified milk fat). *Journal of Molecular Structure, 1153*, 275–281. https://doi.org/10.1016/j.molstruc.2017.09.116.

Utami, P. I., Rahayu, W. S., Nugraha, I., & Rochana, A. N. (2018). Fatty acid analysis of lipid extracted from rats by gas chromatography-mass spectrometry method. *IOP Conference Series: Materials Science and Engineering, 288*(1). https://doi.org/10.1088/1757-899X/288/1/012115.

Vichasilp, C., & Poungchompu, O. (2014). Feasibility of detecting pork adulteration in halal meatballs using near infrared spectroscopy (NIR). *Chiang Mai University Journal of Natural Sciences, 13*(1), 497–507. https://doi.org/10.12982/cmujns.2014.0052.

Wang, D., Wang, L., Xue, C., Han, Y., Li, H., Geng, J., & Jie, J. (2020). Detection of meat from horse, donkey and their hybrids (mule/hinny) by duplex real-time fluorescent PCR. *PLoS One, 15*(12 December), 1–9. https://doi.org/10.1371/journal.pone.0237077.

Xiao-ying, N., Li-min, S., Fang, D., Zhi-lei, Z., & Yan, Z. (2014). Discrimination of donkey meat by NIR and chemometrics. *Spectroscopy and Spectral Analysis, 10*, 1. http://en.cnki.com.cn/Article_en/CJFDTOTAL-GUAN201410032.htm.

Xu, L., Cai, C. B., Cui, H. F., Ye, Z. H., & Yu, X. P. (2012). *Rapid discrimination of pork in Halal and non-Halal Chinese ham sausages by Fourier transform infrared (FTIR) spectroscopy and chemometrics, 92*, 506–510. https://doi.org/10.1016/j.meatsci.2012.05.019.

Yang, L., Wu, T., Liu, Y., Zou, J., Huang, Y., Babu, S. V., & Lin, L. (2018). Rapid identification of pork adulterated in the beef and mutton by infrared spectroscopy. *Journal of Spectroscopy, 2018*. https://doi.org/10.1155/2018/2413874.

Chapter 6
Identification of Potential Biomarkers of Porcine Gelatin

Nur Azira Tukiran, Amin Ismail, Haizatul Hadirah Ghazali, and Nurul Azarima Mohd Ali

Abstract This chapter discusses the identification of potential biomarkers of porcine gelatin using peptide mass fingerprinting (PMF) for the development of anti-peptide polyclonal antibodies. The selection of antigens is a prerequisite for the development of enzyme-linked immunosorbent assay (ELISA). A synthetic peptide is one of the prevalent antigens for ELISA development. It enables the produced antibodies to be targeted at small regions of the protein with fine specificity. It has been used to solve food authenticity issues.

Keywords SDS-PAGE · Peptide mass fingerprinting · In-gel tryptic digestion · Polypeptide band · Porcine gelatin

6.1 Introduction

The enzyme-linked immunosorbent assay (ELISA) is an immunological method that integrates a labeled enzyme (a protein that converts the substrate into a color product) to indicate the presence of antigen-antibody interaction in a sample. This assay has been widely employed in assessing food authenticity due to its rapidity, specificity, sensitivity, and cost-effectiveness (Asensio et al., 2008). The selection of antigens is a prerequisite for ELISA development. A synthetic peptide is one of the most popular antigens for ELISA development. It enables the produced antibodies to be targeted at small regions of the protein with a fine specificity. It has been used to overcome

N. A. Tukiran (✉)
International Institute for Halal Research and Training (INHART), International Islamic University Malaysia (IIUM), Jln Gombak, Selangor 53100, Malaysia
e-mail: aziratukiran@iium.edu.my

A. Ismail
Faculty of Medicine and Health Sciences, Universiti Putra Malaysia (UPM), UPM Serdang 43400, Malaysia

H. H. Ghazali · N. A. M. Ali
Kulliyyah of Science, International Islamic University Malaysia (IIUM), Kuantan 25200, Malaysia

© The Author(s), under exclusive license to Springer Nature Switzerland AG 2021
A. Amid (ed.), *Multifaceted Protocols in Biotechnology, Volume 2*,
https://doi.org/10.1007/978-3-030-75579-9_6

food authenticity issues (Doi et al., 2009; Kreuz et al., 2012; Reed & Park, 2010; Venien & Levieux, 2005).

Many elements will affect the effectiveness of the use of synthetic peptides to increase antibodies. These include the peptide length, the availability of sequence data, and the predicted secondary structure of the intact protein. There is currently a wide range of predictive algorithms available. It was used to predict the antigenicity, hydrophilicity, and surface probability of a given amino acid sequence.

The Chou-Fasman and Garnier-Robson methods laid the foundation for many predictive algorithms of secondary structure to define structure such as α-helix, β-sheet, turns, and coils (Chou & Fasman, 1978; Garnier & Robson, 1989). Thus, the design of the antigen is one of the most significant steps in the entire ELISA development process. Nevertheless, there is no assurance that the produced anti-peptide antibodies will cross-react with the intact protein from which the sequence is obtained because there is a possibility that the target peptide sequence will be buried in the folded protein (Hancock & O'Reilly, 2005).

6.2 Principle

Recently, the use of synthetic peptide as an antigen is often favored. The peptide was generated by mimicking the selected protein region and coupled to the carrier protein before immunization. Since gelatin is a non-antigenic protein, it would be a good strategy to exploit this approach for the determination of porcine gelatin. The success of developing an ELISA largely relies on the binding specificity of the produced antibodies. Thus, the discovery of an appropriate antigen that indicative of both raw and processed products would be crucial. A protein that exhibits exceptional resistance to heat or technological processes will be a good biomarker. It is important to note that identification of species source requires a species-specific sequence of the protein concerned. Preliminary studies on the identification of protein biomarkers as well as species-specific sequences are therefore necessary.

6.2.1 Objective of Experiment

This experiment aims to identify the potential marker peptides of porcine gelatin by peptide mass fingerprinting (PMF). In the present study, sodium dodecyl sulfate polyacrylamide gel electrophoresis (SDS-PAGE) was employed to separate the denatured protein of porcine gelatin. The prominent band was then excised and in-gel digested with trypsin before tandem mass spectrometry (MS/MS) analysis. Subsequently, the identified marker peptides were used for incorporation into the ELISA. The flowchart of the experiment is shown in Fig. 6.1.

Fig. 6.1 Flowchart of
experiment

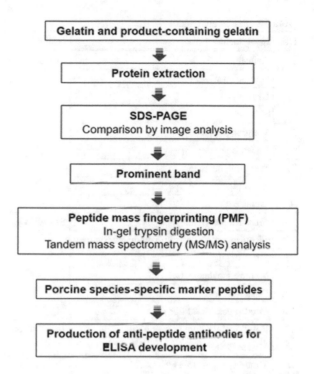

6.3   Materials

The list of materials for sample preparation, SDS-PAGE, in-gel tryptic digestion of
polypeptide bands, and protein identification is listed in Tables 6.1, 6.2, 6.3 and 6.4.

Table 6.1   List of samples

No.	Sample	Manufacturer
1.	Porcine skin gelatin (Type A, 300 Bloom)	Sigma-Aldrich (St. Louis, MO, USA)

Table 6.2   List of consumable

No.	Consumable	Usage
1.	Centrifuge tube (1 ml, 15 ml, 50 ml)	Sample preparation, heat treatment, SDS-PAGE, in-gel tryptic digestion of polypeptide bands, protein identification

Table 6.3 List of equipment

No.	Equipment	Usage
1.	Beaker	Sample preparation
2.	Microcentrifuge	
3.	Chiller (4 °C)	
4.	Scissor	
5.	Weighing balance	
6.	Centrifuge	
7.	Weighing balance	
8.	Micropipettes and tips	
9.	Water bath	
10.	Measuring cylinder	SDS-PAGE
11.	Electrophoresis apparatus with accessories	
12.	Fume hood	
13.	Weighing balance	
14.	Densitometer	
15.	Micropipettes and tips	
16.	Rotational vacuum concentrator	In-gel tryptic digestion of polypeptide bands
17.	Micropipettes and tips	
18.	Time-of-Flight (Q-TOF) LC/MS	Protein identification
19.	Micropipettes and tips	

6.4 Methodology

6.4.1 Sample Preparation

Porcine gelatin was mixed with deionized water. The swelled sample was then heated in a water bath until it was completely dissolved. Subsequently, four volumes of cold acetone were added to the mixture. The mixture was then centrifuged at 14,000 rpm for 10 min to separate the supernatant fluid from the precipitate. The supernatant was discarded from the precipitate. The precipitate was then dissolved in a sample buffer consisting of 8 M urea, 20% v/v SDS, 10 mM ETDA, 0.5 M Tris–HCl, 1.114 g/mL, 2-mercaptoethanol, 0.1% v/v glycerol and 0.05% w/v bromophenol blue prior to SDS-PAGE analysis.

Table 6.4 List of chemicals

No.	Chemicals	Usage
1.	Sodium dodecly sulphate (SDS)	SDS-PAGE
2.	Glycine	
3.	Tris-base	
4.	2-mercaptoethanol	
5.	Coomassie brilliant blue R250	
6.	N,N,N',N'-tetramethylethylenediamine (TEMED)	
7.	Glycerol	
8.	Ammonium persulfate	
9.	Ethylcnediaminetetraacetic acid disodium salt (EDTA)	
10.	Urea	
11.	Acrylamide/bis-aciylamide 30%	
12.	1.5 M Tris-HCl buffer pH 8.8	
13.	0.5 M Tris-HCl buffer pH 6.8	
14.	Bromophenol blue	
15.	Methanol	
16.	Acetic acid	
17.	Silver staining	
18.	Formic acid	In-gel tryptic digestion of polypeptide bands
19.	Iodoacetamide	
20.	Trypsin	
21.	Ammonium bicarbonate	
22.	Acetonitrile (ACN)	
23.	Dithiothreitol (DTT)	
24.	Iodoacetamide (IAA)	

6.4.2   SDS-PAGE

The SDS-PAGE was conducted on a slab gel consisting of 6% resolving gel and 4% stacking gel at 80 V for 2 h. The gels were stained in 0.05% (w/v) of Coomassie Brilliant Blue R-250, 15% (v/v) of methanol and 5% (v/v) of acetic acid at room temperature, and destained in 10% (v/v) of acetic acid and 30% (v/v) of methanol until the background were cleared. A densitometer was then used to scan and analyze the stained gels.

6.4.3 Peptide Mass Fingerprinting (PMF)

6.4.3.1 In-Gel Tryptic Digestion of Polypeptide Bands

Stained bands of interest were excised and destained in 50 mM of ammonium bicarbonate/50% acetonitrile (ACN) 1:1 (v/v). The bands were then reduced in 150 µL of a solution containing 10 mM dithiothreitol (DTT)/100 mM ammonium bicarbonate at 60 °C for 30 min and alkylated in 150 µL of 55 mM iodoacetamide (IAA)/100 mM ammonium bicarbonate at room temperature in the dark for 20 min. Subsequently, the gel pieces were washed twice in 50% ACN/100 mM ammonium bicarbonate and followed by dehydration of the gel pieces with 100% ACN and drying in a rotational vacuum concentrator (CHRiST RVC 2–18, Germany). The dried gel pieces were then rehydrated with 25 µL of 6 ng/µL trypsin in 50 mM ammonium bicarbonate buffer and digested at 37 °C overnight. After this, tryptic peptides were extracted using 50% ACN for 15 min, followed by 100% ACN for 15 min. The extracted solutions were then pooled into a single tube and dried in a vacuum centrifuge and followed by solubilization with 100 µL of 1% formic acid/2% ACN.

6.4.3.2 Protein Identification

The resulting peptides were analyzed using Agilent 6520 Accurate-Mass Quadrupole Time-of-Flight (Q-TOF) LC/MS (Agilent Technologies Inç., Santa Clara, CA, U.S). Data analysis was carried out using MassHunter Qualitative Analysis Software (Agilent Technologies Inc. U.S). MS/MS data were analyzed using SpectrumMill Software (Agilent Technologies Inc. U.S) against the NCBI (Fig. 6.2). Sequence similarity between species was determined using the Basic Local Alignment Search Tool (BLAST) (http://blast.ncbi.nlm.nih.gov/).

6.5 Results and Discussion

6.5.1 Protein Identification of Targeted Polypeptide Band

Mass spectrometry was used to analyze the prominent 125 kDa band. Based on a PMF analysis using LC-MS/MS, porcine collagen alpha-2(I) chain was identified. As shown in Table 6.5, a total of 10 peptide sequences containing 187 amino acid residues were obtained, corresponding to the sequence of collagen alpha-2(I) chain-like (*Sus scrofa*).

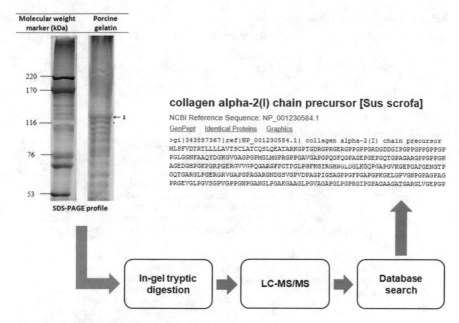

Fig. 6.2 Peptide mass fingerprinting (PMF)

Table 6.5 Ten peptide sequences matched to the tryptic peptide mass fingerprint of the collagen alpha-2(I) chain-like (*Sus scrofa*)[a]

Start-end position	m/z Measured (Da)	MH+ Matched (Da)	Sequence
451–464	637.3136	1241.627	GFpGSpGNVGPAGK
366–375	434.7356	852.469	VGApGPAGAR
328–342	620.3257	1223.649	GIpGPAGAAGATGAR
574–588	727.3777	1421.754	GIpGEFGLpGPAGPR
975–996	504.0081	1997.006	HGNRGEpGPAGSVGPAGAVGPR
100–132	996.4642	2923.408	GPpGAVGAPGpQGFQGPAGEpGEPGQTGpAGAR
883–906	1066.0637	2083.130	GLpGVAGSVGEpGPLGIAGpPGAR
795–815	929.4701	1793.955	IGppGPSGISGPpGPpGPAGK
310–327	781.9188	1514.844	GAAGLpGVAGApGLPGpR
816–830	388.4620	1550.815	EGLRGPRGDQGPVGR

[a]NCBI accession number: gi|343887367; theoretical mass 129321.8

Table 6.6 List of peptide immunogens

Peptide	Sequence	Peptide length	Amino acid position	Protein identification
1.	GFPGSPGNVGPAGK	14	451–464	Collagen, type I, alpha 2 [Sus scrofa]
2.	GIPGEFGLPGPAGPR	15	574–588	

Table 6.7 Multiple alignments of peptides

Peptide	Peptide length	Species	Accession number		Multiple alignments	
1	14	Porcine	NP_001230584	394	GPPGLRGNPGSRGLPGADGRAGVMGPPGSRGPTGPAGVRGPNGDSGRPGEPGLMGPRGFPGSPGNVGPAGKEGPAGLPGI	473
		Donkey	ACM24775	392	GPPGLRGSPGSRGLPGADGRAGVMGPAGSRGATGPAGVRGPNGDSGRPGEPGLMGPRGFPGSPGNIGPAGKEGPVGLPGI	471
		Bovine	NP_776945	392	GPDGLRGNPGSRGLPGADGRAGVMGPAGSRGATGPAGVRGPNGDSGRPGEPGLMGPRGFPGSPGNIGPAGKEGPVGLPGI	471
		Sheep	XP_004007775	392	GPPGLRGNPGSRGLPGADGRAGVMGPAGSRGATGPAGVRGPNGDSGRPGEPGLMGPRGFPGSPGNIGPAGKEGPAGLPGI	471
		Goat	XP_005678993	392	GPPGLRGNPGSRGLPGADGRAGVMGPAGSRGATGPAGVRGPNGDSGRPGEPGLMGPRGFPGSPGNIGPAGKEGPAGLPGI	471
		Chicken	NP_001073182	393	GPAGLRGVPGSRGLPGADGRAGVMGPAGNRGASGPVGAKGPNGDAGRPGEPGLMGPRGLPGQPGSPGPAGKEGPVGFPGA	472
		Fish	XP_006636081	384	GNRGERGVTGSRGLPGLEGRAGPMGMPGARGATGAGGPRGPPGDGGRPGEPGQTGARGLPGSPGSSGPAGKEGPAGAPGQ	463
					451 – 464*	
2	15	Porcine	NP_001230584	554	FQGLPGPAGTAGEVGKPGERGIPGEFGLPGPAGPRGERGPPGESGAAGPAGPIGSRGPSGPPGPDGNKGEPGVLGAPGTA	633
		Donkey	ACM24775	552	FQGLPGPAGTAGEVGKPGERGLPGEFGLPGPAGARGERGPPGESGAAGPAGPIGSRGPSGPPGPDGNKGEPGVLGAPGTA	631
		Bovine	NP_776945	552	FQGLPGPAGTAGEAGKPGERGIPGEFGLPGPAGARGERGPPGESGAAGPTGPIGSRGPSGPPGPDGNKGEPGVVGAPGTA	631
		Sheep	XP_004007775	552	FQGLPGPAGTAGEAGKPGERGIPGEFGLPGPAGARGERGPPGESGAAGPTGPIGSRGPSGPPGPDGNKGEPGVVGAPGTA	631
		Goat	XP_005678993	552	FQGLPGPAGTAGEAGKPGERGIPGEFGLPGPAGARGERGPPGESGAAGPTGPIGSRGPSGPPGPDGNKGEPGVVGAPGTA	631
		Chicken	NP_001073182	553	FQGLPGPSGPAGEAGKPGERGLHGEFGVPGPAGPRGERGLPGESGAVGPAGPIGSRGPSGPPGPDGNKGEPGNVGPAGAP	632
		Fish	XP_006636081	544	FQGLPGPAGPAGETGKAGDRGIPGDAGLPGPAGPRGERGNPGPAGSQGPAGPIGARGASGTPGPDGSKGEPGVAGTVGAA	623
					574 – 588*	

*Amino acid position

6.5.2 *Potential Marker Peptides as Immunogens*

Antibodies were raised against porcine species-specific amino acid sequences of the collagen alpha-2(I) chains (Table 6.6). Alignment-based sequence analysis using BLAST showed that the porcine amino acid sequence was species-specific when compared with bovine, donkey, sheep, goat, chicken, and fish (Table 6.7). Venien & Levieux (2005) used the same approach to detect bovine gelatin in porcine gelatin where anti-peptide antibodies are used against the bovine collagen alpha-1(I) chain. Besides, anti-peptide antibodies for mammalian gelatin quantification have also been used as capture antibodies in the sandwich ELISA (Doi et al., 2009). Due to the poor immunogenicity of gelatin, this approach can offer an alternative to other authentication methods.

6.6 Conclusion

The selection of a suitable antigen is a prerequisite for ELISA development. The prominent polypeptide band (125 kDa) of SDS-PAGE of porcine gelatin was identified as the collagen alpha-2(I) chain. Two amino acid sequences, GFPGSPGN-VGPAGK (Peptide 1) and GIPGEFGLPGPAGPR (Peptide 2), of porcine species-specific, were selected for raising polyclonal antibodies to be introduced into the ELISA. This approach has been effective in developing ELISA for porcine gelatin detection.

Acknowledgements The authors would like to acknowledge the Ministry of Science, Technology, and Innovation (MOSTI), Malaysia, for awarding funds via the Science Fund (Fund no. 02-01-04-SF1447).

References

Asensio, L., Gonzalez, I., Garcia, T., & Martin, R. (2008). Determination of food authenticity by enzyme-linked immunosorbent assay (ELISA). *Food Control, 19*, 1–8.

Chou, P. Y., & Fasman, G. D. (1978). Prediction of the secondary structure of proteins from their amino acid sequence. *Advances in Enzymology and Related Areas of Molecular Biology, 47*, 45–47.

Doi, H., Watenabe, E., Shibata, H., & Tanabe, S. (2009). A reliable enzyme linked immonosorbent assay for the determination of bovine and porcine gelatin in processed foods. *Journal of Agricultural and Food Chemistry, 57*, 1721–1726.

Garnier, J., & Robson, B. (1989). In prediction of protein structure and principles of protein conformation, G. Fasman (Eds.), Ch. 10: 417–465. Plenum.

Hancock, D. C., & O'Reilly, N. J. (2005). Synthetic peptides as antigens for antibody production. In *Methods in molecular biology* (Vol. 295), Immunochemical Protocols, Third Edition. Edited by R. Burns © Humana Press Inc.

Kreuz, G., Zagon, J., Broll, H., Bernhardt, C., Linke, B., & Lampen, A. (2012). Immunological detection of osteocalcin in meat and bone meal: A novel heat stable marker for the investigation of illegal feed adulteration. *Food Additive and Contaminants: Part A, 29*, 716–726.

Reed, Z. H., & Park, J. W. (2010). Quantification of Alaska Pollock surimi in prepared crabstick by competitive ELISA using a myosin light chain 1 specific peptide. *Food Chemistry, 123*, 196–201.

Venien, A., & Levieux, D. (2005). Differentiation of bovine from porcine gelatines using polyclonal anti- peptide antibodies in direct and competitive indirect ELISA. *Journal of Pharmaceutical and Biomedical Analysis, 39*, 418–424.

Chapter 7
Gamma Ray Mutagenesis on Bacteria Isolated from Shrimp Farm Mud for Microbial Fuel Cell Enhancement and Degradation of Organic Waste

Ayoub Ahmed Ali, Azura Amid ⓘD, and Azhar Muhamad

Abstract This chapter discusses the use of prokaryotic microorganisms to produce electrical bioenergy from a wide range of organic substrates in microbial fuel cells (MFCs). MFCs offer promising sustainable energy production and at the same time, simultaneous degradation of organic waste in the substrate. Active microorganisms capable of producing electricity bypassing the electron to the electrode are called electrochemically active bacteria. The study identified a method to obtain an optimum dose to increase the bacterial potential using the one factor at time (OFAT) method. The 63 Gy gamma dose irradiation increased the cell voltage to 280 mV with 33% of chemical oxygen demand (COD) removal while the maximum voltage of the wild strain was 154 mV with 55.7% of COD removal. The successful effect of the gamma radiation dose on the increase of the MFC's bioelectricity and organic matter removal indicates that gamma rays are a way to boost the ability of the electrically active bacteria.

Keywords Gamma ray · Microbial fuel cell · Mutagenesis

7.1 Introduction

All living organisms need energy, and there are two main sources of energy: chemical and light. Organisms that use light as a source of energy are called phototrophs, while others are chemotropic. Microbial fuel cell (MFC), as the French claim, "living

A. A. Ali
Kulliyyah of Engineering, International Islamic University Malaysia (IIUM),
Kuala Lumpur, Malaysia

A. Amid (✉)
International Institute for Halal Research and Training, International Islamic University Malaysia,
Jalan Gombak, Kuala Lumpur, Malaysia
e-mail: azuraamid@iium.edu.my

A. Muhamad
Agensi Nuklear Malaysia, Kajang, Selangor, Malaysia

© The Author(s), under exclusive license to Springer Nature Switzerland AG 2021
A. Amid (ed.), *Multifaceted Protocols in Biotechnology, Volume 2*,
https://doi.org/10.1007/978-3-030-75579-9_7

battery" generates electricity from a chemotrophic bacterium that transforms chemical energy (organic matter present in wastewater) into electrical energy. An organism that drives electrical current is called electrochemically active bacteria, electricigens, and anode respiring bacteria (Logan, 2009).

7.1.1   Principle: Microbial Fuel Cell (MFC)

The standard configuration of the MFC system consists of two interconnected chambers. The anode chamber is filled with a medium in which the living catalyst breaks down the organic or inorganic materials of the media and produces an electron from its metabolism on the surface of the electrode. On the other hand, the cathode chamber contains electron acceptors where several electron acceptors have been used, e.g. oxygen (O_2), phosphate buffer solution (PBS), ferricyanide ($K_3Fe[CN]_6$) and nitrate (NO_3) (Wei et al., 2012; Wen et al., 2010; Zhao et al., 2006). The two chambers are separated by a membrane which not only separates the solutions of each compartment but preserves the electro-neutrality of the system by diffusing the proton from the anode to the cathode. The reactions in anode and cathode are described from the following equations.

$$C_6H_{12}O_6 + 6H_2O \rightarrow 6CO_2 + 24H^+ + 24e^- \tag{7.1}$$

$$6O_2 + 24H^+ + 24e^- \rightarrow 12H_2O \tag{7.2}$$

7.1.2   Principle: Mutation

A mutation is a modification of the genomic arrangement that can alter or disable the functionality of a cell. The disorder may occur at the nucleotide level (nucleotide deletion, alteration, and replacement) as well as at the chromosomal level (chromosome rearrangement or duplication) or other levels. Traditionally, mutations are classified into three categories (single gene disorder, chromosome disorder, and multifactorial disorder). Single gene disorder is an error that occurred in the DNA sequence of the gene. It is categorized as somatic (non-reproductive cell) and germline (reproductive cell). The somatic disorder may be so-called autosomal recessive in the recessive allele or maybe on the autosomal dominant. Where the germline is either on the X chromosome (X-linked recessive and dominant disorder) or Y chromosome (holandric disorder) (Mahdieh & Rabbani, 2013). Thus, the basis of the category may or may not appear in the phenotype of the offspring. The mutation occurs spontaneously or is induced by an agent named mutagens such as chemical, radiation, viruses, diet, or lifestyle. A spontaneous mutation is natural; it appears without any induction agent and is the result of a mispairing nucleotide (added or omitted) during the DNA replication phase that is contrary to the mutation induced by a mutagen that can damage the DNA and result in nucleotide mispairing (Watford & Warrington, 2017).

7.1.2.1 Chemical Mutagens

Due to the incapability of a cell to produce its thymidine or the presence of thymidine inhibitor in the media because it is replaced by 5-bromodeoxyuridine (BUdR), an unstable molecule converted enzymatically by 5-bromouracil (Fishbein, 1970). It makes the same mistake when pairing with guanine instead of adenine during DNA replication. The transition of A-T to G-C is thought to be due to a higher electronegativity of the bromine atom compared to the methyl group of thymidine, rendering the ring weaker in electrons and favoring guanine pairing. Additionally, Fishbein (1970) emphasizes that the transition does not stabilize in G-C pairing, but an occasional error leads the pairing BUdR to adenine. Moreover, proflavin, acridine orange, and ethidium bromide are mutagens that can slip between the stacked nitrogen bases at the core of the DNA double helix. In this position, these agents can add or delete nucleotides in the sequence (Habibi & Pezeshki, 2013). Furthermore, some chemicals, such as nitrous acid, can chemically modify the base so that it appears like a different base (Jena, 2012).

7.1.2.2 Ultraviolet Irradiation

Ultraviolet in a wavelength between 200 and 280 nm incorporates the formation of the bond between adjacent pyrimidine amino acids, commonly referred to as thymine dimers for DNA and uracil dimers for RNA (Cutler & Zimmerman, 2011). They also reported that these dimers set up the blockade of DNA replication as well as the transcription.

7.1.2.3 Radioactive Particle

There are five forms of ionizing radiation: alpha, beta, neutron, gamma, and X-rays. They caused an unstable atom with an excess of mass and/or energy. For the atom to stabilize, it must free that extra energy and/or mass in the form of radiation. Alpha and beta rays are positively and negatively charged, while gamma rays are the result of the Compton effect or Compton scattering (Parks, 2015). Compton scattering is the quantum theory where a scattering photon collides with an atom. Basically, the energy of the incoming photon is partly absorbed by an outer shell electron, where the electron is not only knocked out from the atom, but a scattering photon is also released (Parks, 2015). Generally, radiation exposure causes a change in the DNA structure and the formation of a free radical that cause the excitation, ionization, and breakage of molecules.

Henceforth, irradiation can inhibit the growth of the living organism or lead to the apparition of mutation that can potentially change the metabolism pathways and morphology of the organism. The irradiation effect depends on the type of radiation, duration of the exposure, the distance between the sample and the source, and the species type (Min et al., 2003; Quillardet et al., 1989). Moreover, the hormesis hypothesis suggests a beneficial effect of low-dose radiation on the microorganism, increases interest in low-dose radiation (Bolsunovsky et al., 2016). Besides, a considerable amount of literature has been published on the effects of ionizing radiation on the viability of bacterial cells, inducing an SOS response; response to DNA damage in which the cell cycle is stopped and initiates DNA repair and mutagenesis in bacterial cells. The comparison of three types of ionizing radiation (alpha, gamma, and neutron radiation) using SOS Chromo test showed a minimum inducing doses (MID) of 2.5 Gy for alpha and neutrons ionizing, and around 5 Gy for gamma rays, with the highest inducing dose, was 100 Gy for neutrons and 200 Gy for gamma and alpha particles (Quillardet et al., 1989). Since the minimum inducing dose were approximately similar for the 3 types of radiation Quillardet, the authors of that investigation stated that the standard Ames test procedure measured 1–2 Gy dose as the lowest mutation doubling dose for gamma rays and concluded that the value of SOS chromotest and Ames test were logically comparable.

Moreover, Min and co-workers (Min et al., 2003) stated that the minimum dose detects by *Escherichia coli* is 1.5 Gy, whereas the maximum response was 200 Gy. The amount of emitted bioluminescence increased proportionally with the gamma-ray doses, which caused DNA damage response in a range of "1–50 Gy". Since the maximum gamma irradiation dose that inhibits bacteria is 200 Gy and the minimum detectable DNA damage dose is 1.5 Gy, the gamma irradiation dose of this study was 25, 63, and 100 Gy.

7.1.3 Objective of Experiment

The objective of this procedure is to increase the MFC performance through gamma irradiation on isolated bacteria.

7.2 Materials

Tables 7.1, 7.2 and 7.3 show the materials used in this experiment.

Table 7.1 Consumable used in this experiment

No.	Consumable	Usage
1	Petri dishes	Bacteria growth
2	Inoculation tube	Bacteria growth
3	Cuvette	Optical density reading
4	COD digestion vial	COD measurement
5	Pipette tips	Liquid handling
6	Micro centrifuge tube	Storage

Table 7.2 Equipment used in this experiment

No.	Equipment	Usage
1	U-tube MFC	Isolation of electrochemically active bacteria
2	Multimeter	Electricity production measurement
3	Spectrophotometer	Bacteria growth measurement
4	Digital multimeter	Resistance and voltage measurement
5	COD reactor	Chemical Oxygen Demand measurement

Table 7.3 Chemicals used in this experiment

No.	Chemicals	Usage
1	LB broth	Bacteria growth

7.3   Methodology

7.3.1   Bacterial Radiation

The identified bacteria were sent to the Malaysian Nuclear Agency, Bangi, Selangor for mutation by gamma radiation. Serial dilution of the LB broth containing the wild-type strain was performed. The reason for the dilution was to cultivate ten colonies in each petri dish to which the radiation was to be applied. Since the high dose will cause inhibition of the bacterium as well as the low dose, the mutation is caused. Also, gamma doses greater than or equal to 200 Gy cause bacterial inhibition (Bolsunovsky et al., 2016). Therefore, the dose of gamma radiation in this study was 25, 63, and 100 Gy.

7.3.1.1   One Factor at a Time (OFAT) Optimization of Gamma Radiation

The isolated *Bacillus cereus* strain cc-1 was irradiated with gamma radiation by the Malaysian nuclear agency. Electrochemically active bacteria were exposed to three different doses (25, 63, 100 Gy) of gamma rays. Tenfold dilution series were performed in order to get a solid plate containing approximately ten colonies of *Bacillus cereus* strain cc-1. Optimization studies on the increase in electricity production of electrochemically active bacteria of MFC using gamma radiation were conducted based on one factor (gamma radiation dose) that has a positive significant effect on the enhancement of cell potential.

Optimization was carried out using One-Factor at Time (OFAT) in order to determine the relationship between the factor and responses (voltage, mV; time, h). The high and low levels of the values of the actual variables were 100 and 25 Gy, respectively. Cell voltage and time were taken as the response to the experimental design. The experiment was conducted in the same reactor (U-tube MFC) with the wild strain containing all its physical properties, and then 1 mL of LB broth containing the mutant strain of 0.1 OD was added. Electricity production was recorded in 3 days using a multimeter (Fluke 179) and the analysis of variance (ANOVA) was conducted using the Design Expert Software.

7.3.2   Bacterial Growth

Bacterial growth was measured by using a spectrophotometer as the selected wavelength of the isolated bacteria was not known to be 600 nm (Fakhirruddin et al., 2018). After the cell was inoculated in Luria broth, 1 mL of the cell-containing nutrient broth was deducted from the inoculation tube and poured into a 1 mL cuvette for reading the optical density of the sample. This process continued until 14 h after inoculation, with an interval of 1 h for the first 6 h and 30 min for the remaining 8 h.

7.3.3   MFC Power Measurement

After constructing an MFC device with an external resistance of 1000 Ohms (refer to Sect. 7.2.1), the voltage generation of the MFC was measured with a digital multimeter. As described in Sect. 7.2.1, the voltage was measured directly in the function of time where the current and the power density were calculated using Eqs. (7.1–7.3)

$$A = V/R \qquad (7.1)$$

where A is the intensity of the current, V is the voltage, and R is the resistance.

The equation of the current density and the power density is written as follows:

$$Am = A/(a) \qquad (7.2)$$

$$Pd = V * Am \qquad (7.3)$$

where Am is the current density (mA/cm^{-2}), Pd is the power density (mW/cm^{-2}), and a is the surface area of the anodic electrode (cm^2).

7.3.4   Total Chemical Oxygen Demand (COD)

Chemical Oxygen demand (COD) is a test that determines the amount of organic compounds in wastewater. If the COD value is very high, this means that there is a high amount of organic matter in the sample. A box of Hach® high range (HR) COD digestion vials was used for the COD test in this research. The vials were all labeled and all TCODs were measured at the same time by using one blank.

The sample was well mixed and 1 mL was pipetted in a 100 mL tube, which was then diluted to 99 mL with distilled water. A 2 mL diluted sample was transferred to a COD vial and a blank was made using 2 mL of distilled water instead of a sample. The COD vials were put and left for 2 h in the COD reactor (Hach DRB200 Digital Reactor Block for TNTplus, Thomas Scientific, US) preheated at 150 °C for 20 min. After 2 h heating, the vials were placed on a wooden rack for 10 min to cool down in the ambient temperature. Finally, the reading was done by using DR 5000™ UV-Vis Laboratory Spectrophotometer (Hach company, US).

7.4   Results and Discussion

7.4.1   Suitable Gamma Rays Dose by OFAT

Design Expert 7.0. (2021 East Hennepin Ave., Suite 480 Minneapolis, MN 55413) was used for the design, analysis, and optimization, where the variable was the gamma radiation dose (25, 63, and 100 Gy). It was the only factor used as a parameter in order to study its effect in response to an increase in electricity performance by mutating the electrochemically active bacteria. The responses were voltage (Y1) and time (Y2).

A total of five treatments was designed to search for the optimum dose (Gy) using the statistical software of DOE applying OFAT, Table 7.4 illustrates the experimental data of OFAT with the two responses. The highest voltage generation with the shortest time was 280 mV in 20 h achieved by the dose of 63 Gy, while the lowest voltage potential was 29.1 mV in 55 h produced by the strain radiated with the lowest dose (25 Gy) (Table 7.4).

The corresponding model graph for the voltage (a) and the time (b) was used in Fig. 7.1 to examine and clearly understand the relationship between the factor (gamma radiation dose) and the two responses. The objective of this model is to increase the voltage generation and reduce time. Therefore, in Fig. 7.1a, a parabolic concave curve is seen towards the maximum point, where the lowest radiation dose

Table 7.4 Experimental data of OFAT with the result

Run	Factor 1: Dose (Gy)	Response 1: Voltage (mV)	Response 2: Time (h)
1	100	44.8	72
2	25	−29.1	55
3	100	44.8	72
4	25	−29.1	55
5	63	280	20

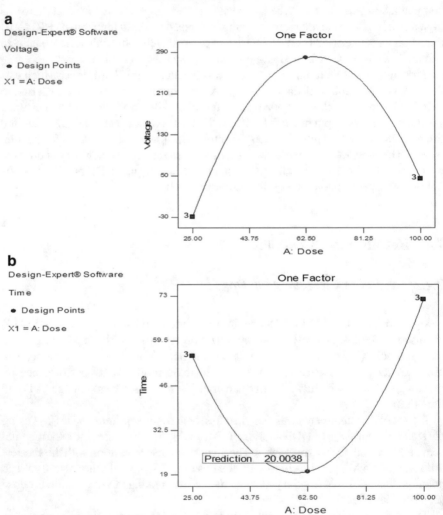

Fig. 7.1 Response curve, **a** voltage over dose; **b** time over dose

and the highest gamma dose caused a considerable drop in cell voltage generation. Thus, the optimal dose that increased the potential voltage was 63 Gy with a voltage of 280 mV. Moreover, Fig. 7.1b demonstrates that both the highest dose and the lowest dose of 100 and 25 Gy, respectively, increased the time to reach the maximum electricity generation, when a convex curve towards the minimum point was observed. Similar to the first response—voltage, the optimum dose that minimized the time of MFC performance was 63 Gy.

7.4.2 Formation of Biofilm in the Anolyte

Figure 7.2 presents the scanning electron microscopy (SEM) images, which show the results of the biofilm formation on the anode electrode. The sample used for the analysis was a carbon cloth, which was obtained from the electrochemical bacteria after being exposed by the optimization dose of the gamma radiation. An electrode sample of 10 × 10 mm was taken from each experiment using a radiated strain and a wild strain, where an unused sterile carbon cloth was used as a reference. All samples undergo the same preparation procedure and the image shown in Fig. 7.2 was taken with a magnification of ×1500.

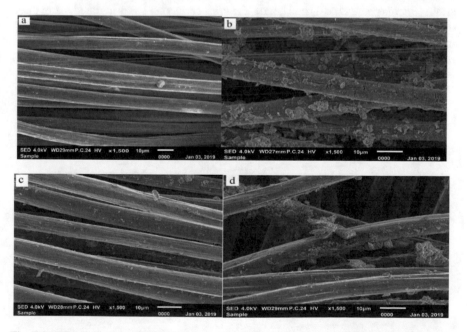

Fig. 7.2 SEM image of carbon cloth anode after 3 days electricity production in U-tube MFC; **a** control ×1500, **b** *Bacillus cereus* strain cc-1, **c** *Bacillus cereus* mutant 25 Gy, **d** *Bacillus cereus* mutant 63 Gy, **e** *Bacillus cereus* mutant 100 Gy

The surface of the control carbon cloth was uncovered of bacteria or biofilm, whereas the surface of the electrode sample from the other test had a deposition with different magnitude of layers. The magnitude of deposition was great on the sample from gamma radiated strain of 63 Gy and the wild strain, and moderate on the sample from 100 Gy. Besides, the surface of the carbon cloth had an insignificant deposition of bacteria for the assay sample that used the strain radiated with 25 Gy dose. This deposition shows the colonization of biofilm cells and their exopolysaccharide (EPS) on the surface of the carbon cloth and in fact, these images are consistent with the result obtained from the optimization process.

7.4.3 Electricity Performance

All experiments have been carried out in triplicates. The electrochemically active bacterium identified was subjected to gamma radiation. It had previously been reported that ionizing radiation induces mutagenic damage to cellular DNA, while 200 Gy induced the maximum gamma dose, causing a 90% bacterial inhibition (Min et al., 2003). Also, the hypothesis of hormesis (Bolsunovsky et al., 2016) suggested that low ionizing radiation (less than 200 Gy) induces a beneficial muta- tion. This experiment was aimed to increase the electricity production of the wild strain (*Bacillus cereus* cc-1) and hence, *Bacillus cereus* cc-1 was mutated using three different doses of gamma radiation (25, 63, and 100 Gy). Therefore, the following information in Fig. 7.3 indicates the generation of bacterial voltage after gamma radiation. Bacteria emitted with 63 Gy radiation produced the highest voltage (280

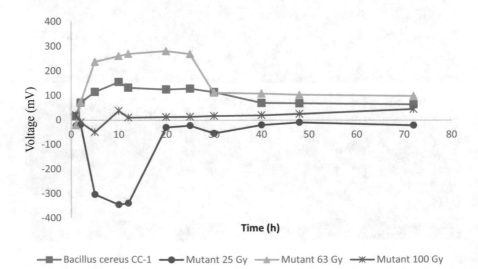

Fig. 7.3 Comparison of cell potential between the three mutant strain and the wild strain of *Bacillus cereus cc-1*

mV) in 20 h and the 25 Gy-irradiated strain showed reverse polarity whereas the highest dose of the experiment (100 Gy) induced a current production loss.

Reverse voltage phenomena are popular in hydrogen fuel cells when one cell of the fuel stack runs out of food (Lee et al., 2006). The investigation by Aelterman et al. (2006) is one of the pioneer studies that observed a reversal of polarization in MFCs, but they believed without experiment that it was due to food starvation. Therefore, Oh and Logan (2007) studied a series of three assays with two cells of MFC stack each where in the first experiment, both cells were inoculated with electrochemically active bacteria and fed perfectly, whereas in the second experiment, both cells were inoculated with electrochemical bacteria but one cell was intentionally deprived and fed the other cell, the third test was similar to the first experiment except one cell was free of the electrochemical bacteria. After monitoring the potential voltage of the stack MFC, the starved bacteria of the second assay and the bacteria-free MFC cell of the third experiment produced a potential reversal. Thus, the lack of food, as well as the lack of bacteria, may therefore cause a reversal voltage.

Moreover, the result from the SEM image of Fig. 7.2 shows that the sample from the test that used the strain irradiated with 25 Gy does not develop a biofilm formation, so either the radiation will cause a loss of metabolism pathway that used mud and water from the shrimp pond as a source of energy, or the bacteria have lost the ability to form a biofilm due to a mutation induced by gamma radiation. To prove these hypotheses, it was important to measure the removal of organic matter. Furthermore, the 100 Gy-dose mutant strain has developed a small layer of biofilm as well as a low electricity production compared to the 63 Gy-irradiated mutants that produced the maximum cell potential with a very thick layer of biofilm on the electrode surface. Henceforth, there is a positive direct correlation between biofilm formation and electricity generation in this study. Additionally, it was reported that the thicker biofilm-forming *Geobacter sulfurreducens* produces higher electricity generation (Sun et al., 2016). Moreover, some bacteria cannot produce electricity during the planktonic phase but can transfer electrons after biofilm formation (Watson & Logan, 2010). Therefore, based on the consequence of SEM and the literature of the reverse polarity induced by the radiation dose can be explained by the loss of bacterial biofilm formation.

7.4.4 The Effect of COD Removal and Strain Relations

COD is used to measure the amount of organic compounds present in the MFC substrate (water and mud from shrimp farm). Figure 7.4 shows the COD of the substrate containing bacteria inoculum (without culture, mix culture, pure culture, and mutants). The COD reading was observed after 3 days in the MFC operation. Mixed culture eliminates more COD than pure strain (55.7 and 65%, respectively). Similarly, Fatemi et al. (2012) found that mixed culture achieves a COD removal efficiency of 75%, whereas pure culture only reduces 54% of organic matter. Additionally, Nimje et al. (2012) also reported that the mixed culture removes more COD

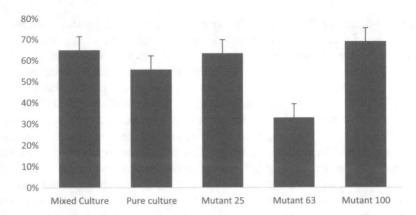

Fig. 7.4 Comparison of the Total COD removal between the mixed culture, the three mutant of strain CC-1 and the wild strain of CC-1

than pure oxygen. In both studies, they concluded that the methanogenic bacteria and denitrifying bacteria present in the mixed culture absorb the organic substrate without electricity generation.

Figure 7.4 depicts that the mutants treated with 25 and 100 Gy, despite their loss in electricity production, remove more COD compared to mutants treated with 63 Gy, 33, 63.4, and 69%, respectively. Therefore, with the mutation induced by gamma irradiation, we obtained a mutant that increased electricity production by more than two times and also, two mutant strains that removed a comparable rate of COD removal compared to that of the mixed culture.

7.5 Conclusion

The usage of low-dose gamma radiation has increased the potential to produce bacterial electricity and to remove organic matter. Since chemical mediator was not used to increasing the electricity generation, this method is considered to be eco-friendlier on the sludge and environment.

Acknowledgements The author would like to acknowledge the International Islamic University, Malaysia for awarding fund via Publication RIGS Grant (Grant No. P-RIGS18-065-0065).

References

Aelterman, P., Rabaey, K., Pham, H. T., Boon, N., & Verstraete, W. (2006). Continuous electricity generation at high voltages and currents using stacked microbial fuel cells. *Environment Science and Technology, 40*(10), 3388–3394.

Bolsunovsky, A., Frolova, T., Dementyev, D., & Sinitsyna, O. (2016). Low doses of gamma-radiation induce SOS response and increase mutation frequency in *Escherichia coli* and *Salmonella typhimurium* cells. *Ecotoxicology and Environmental Safety, 134,* 233–238.

Cutler, T. D., & Zimmerman, J. J. (2011). Ultraviolet irradiation and the mechanisms underlying its inactivation of infectious agents. *Animal Health Research Reviews/Conference of Research Workers in Animal Diseases, 12*(1), 15–23.

Fakhirruddin, F., Amid, A., Wan Salim, W. W. A., & Azmi, A. S. (2018). Electricity generation in Microbial Fuel Cell (MFC) by bacterium isolated from rice paddy field soil. *E3S Web of Conferences, 34,* 1–9. https://doi.org/10.1051/e3sconf/20183402036.

Fatemi, S., Ghoreyshi, A. A., Najafpour, G., & Rahimnejad, M. (2012). Bioelectricity generation in mediator—Less microbial fuel cell: Application of pure and mixed cultures. *Iranica J Energy & Environment, 3*(2), 104–108.

Fishbein, L. (1970). *Alkylating agent in chemical mutagens environmental effects on biological systems* (1st ed.). Academic Press.

Habibi, M. B., & Pezeshki, N. P. (2013). Bacterial mutation; Types, mechanisms and mutant detection methods: A review. *European Sc. J, 4*(December), 1857–7881.

Jena, N. R. (2012). DNA damage by reactive species: Mechanisms mutation and repair. *J Biosciences, 37*(3), 503–507.

Lee, H., Song, H., & Kim, J. (2006). Effect of reverse voltage on proton exchange membrane fuel cell performance, pg. 205–208 in 2006 International Forum on Strategic Technology, IFOST. IEEE Xplore.

Logan, B. E. (2009). Exoelectrogenic bacteria that power microbial fuel cells. *Nature Reviews Microbiology, 7,* 375–381.

Mahdieh, N., & Rabbani, B. (2013). An overview of mutation detection methods in genetic disorders. *Iranian J Pediatrics, 23*(4), 375–388.

Min, J., Lee, C. W., & Gu, M. B. (2003). Gamma-radiation dose-rate effects on DNA damage and toxicity in bacterial cells. *Radiation and Environmental Biophysics, 42*(3), 189–192.

Nimje, V. R., Chen, C. Y., Chen, H. R., Chen, C. C., Huang, Y. M., Tseng, M. J., Cheng, K. C., & Chang, Y. F. (2012). Comparative bioelectricity production from various wastewaters in microbial fuel cells using mixed cultures and a pure strain of Shewanella oneidensis. *Bioresource Technology, 104,* 315–323. https://doi.org/10.1016/j.biortech.2011.09.129.

Oh, S., & Logan, B. E. (2007). Voltage reversal during microbial fuel cell stack operation. *Journal of Power Sources,167*(1), 11–17.

Parks, J. E. (2015). *The compton effect-compton scattering and gamma ray spectroscopy.* Knoxville: The University of Tennessee.

Quillardet, P., Frelat, G., Nguyen, V. D., & Hofnung, M. (1989). Detection of ionizing radiations with the SOS chromotest, a bacterial short-term test for genotoxic agents. *Mutation Research/Environmental Mutagenesis and Related Subjects, 216*(5), 251–257.

Sun, D., Chen, J., Huang, H., Liu, W., Ye, Y., & Cheng, S. (2016). The effect of biofilm thickness on electrochemical activity of Geobacter Sulfurreducens. *International Journal of Hydrogen Energy, 41*(37), 16523–16528.

Watford, S., & Warrington, S. J. (2021). *Bacterial DNA mutations.* Treasure Island (FL): StatPearls Publishing.

Watson, V. J., & Logan, B. E. (2010). Power production in MFCs inoculated with Shewanella oneidensis MR-1 or mixed cultures. *Biotechnology and Bioengineering, 105*(3), 489–498. https://doi.org/10.1002/bit.22556.

Wei, L., Han, H., & Shen, J. (2012). Effects of cathodic electron acceptors and potassium ferricyanide concentrations on the performance of microbial fuel cell. *International Journal of Hydrogen Energy, 37*(17), 12980–12986.

Wen, Q., Wu, Y., Zhao, L., & Sun, Q. (2010). Production of electricity from the treatment of continuous Brewery wastewater using a microbial fuel cell. *Fuel, 89*(7), 1381–1385.

Zhao, F., Harnisch, F., Schroder, U., Scholz, F., Bogdanoff, P., & Herrmann, I. (2006). Challenges and constraints of using oxygen cathodes in microbial fuel cells. *Environment Science and Technology, 40*(17), 5193–5199.

Chapter 8
Effect of Temperature on Antibacterial Activity and Fatty Acid Methyl Esters of Carica Papaya Seed Extract

Muhamad Shirwan Abdullah Sani, Jamilah Bakar, Russly Abdul Rahman, and Faridah Abas

Abstract This chapter addresses the antibacterial activity of *Carica papaya* seeds due to their abundance in bioactive compounds and these seeds contain high levels of fatty acid methyl esters (FAMEs). However, no report is available to indicate (1) which FAMEs are potent against pathogens and (2) the effect of temperature on the distribution of FAMEs. Therefore, this study aims to evaluate the effect of temperature against the antibacterial activity of *Carica papaya* seed extract (CPSE) and its FAME profile via extraction of the seeds using methanol and the extract was subjected to test of antibacterial activity against *Salmonella enteritidis, Bacillus cereus, Vibrio vulnificus,* and *Proteus mirabilis.* FAME profiling was done using GC/MS incorporated with principal component analysis (PCA). The CPSE at 5.63 mg/mL was potent against these pathogens at < 40 °C. Although the CPSE was rich with FAMEs, the PCA result had identified individual FAMEs that inhibited the pathogen growth. Palmitic acid (C16:0), palmitoleic acid (C16:1), stearic acid (C18:0), oleic acid (C18:1n9c), and cis-vaccenic acid (C18:1n11c) had strongly inhibited *V. vulnificus* and *P. mirabilis* growths and moderately inhibit *S. enteritidis* growth. To avoid the formation of trans FAMEs, this study also suggested that the CPSE temperature should be held at < 150 °C.

Keywords *Carica papaya* seed · GC/MS · Antibacterial activity · Toxicity · Stability

M. S. A. Sani (✉)
International Institute for Halal Research and Training, International Islamic University, Kuala Lumpur 53100, Malaysia

J. Bakar · R. A. Rahman · F. Abas
Department of Food Technology, Faculty of Food Science and Technology, Universiti Putra Malaysia, Serdang 43400, Selangor, Malaysia

© The Author(s), under exclusive license to Springer Nature Switzerland AG 2021
A. Amid (ed.), *Multifaceted Protocols in Biotechnology, Volume 2*,
https://doi.org/10.1007/978-3-030-75579-9_8

8.1 Introduction

Papaya or *Carica papaya* is a highly commercialized tropical fruit. Ripe *Carica papaya* is mainly eaten as a dessert and papain from the plant's latex is commercialized as a meat tenderizer and is also been used as an enzyme in several enzymatic extraction works (Vasu et al., 2012). Not much attention has been paid to the fruit as a potential source of phytochemicals other than the flesh of the fruit itself. Out of 28 million metric tons of produced papaya worldwide, 5 million metric tons of the seeds were discarded in 2017 (FAOSTAT, 2019). Although *Carica papaya* seeds are discarded from other parts of the world, they are popularly consumed in India, Central Asia, and the Middle Eastern countries. They have been used to marinate meat, substitute for black pepper, and are added to salad dressings due to their spicy flavor (Lim, 2012). Several reports on antibacterial (Sani, 2018), antifungal (Chávez-Quintal et al., 2011) and anticancer (Nakamura et al., 2007) activities of *Carica papaya* seeds obtained from several solvent extracts could be found. Moreover, Sani et al. (2017b) found that the crude *Carica papaya* seed extract (CPSE) was active against *S. enteritidis, V. vulnificus, P. mirabilis, and B. cereus*, but reported potent antibacterial compounds. Since the yield and consumption of *Carica papaya* seeds are very high, it is worth investigating their compositions and potent antibacterial components that potentially inhibit these pathogenic bacteria.

The composition of CPSE was dominated by fatty acid methyl esters (FAMEs) (Sani et al., 2017b). However, there was limited information on which FAMEs had rendered bacterial inhibition. Also, trans FAMEs are of high concern because they have been linked to nutritional health issues (Tao, 2007), especially C18:1n9t (Preedy et al., 2011) while subjecting heat treatment. Cis and trans FAMEs have been stated to be hardly separated due to the same molecular weight except in the use of HP-88 column and GC/MS detection (Albuquerque et al., 2011). Thus, this study was performed in order to (1) investigate the efficacy of antibacterial CPSE and its distribution of FAMEs as heat-affected and (2) identify potential FAMEs that render antibacterial activity in CPSE.

8.2 Methodology

8.2.1 Plant Material

Carica papaya cv. Sekaki fruits were bought from D'Lonek Sdn. Bhd. Organic Farm, Rembau, Negri Sembilan, Malaysia. *Carica papaya* plant, flower, and fruit from this farm were deposited to Herbarium of the Institute of Bioscience, Universiti Putra Malaysia, and a plant voucher, numbered as SK 2368/14, was issued. The seeds of *Carica papaya* were removed from the fruit and treated as described by Sani et al. (2017a). The seeds were thoroughly washed in distilled water, oven-dried at 40 °C for three days, kept in airtight amber bottles, and stored at −20 °C until further analysis.

8.2.2 Extraction of Phytochemicals

Methanol (MeOH) was the solvent used for extraction. Dried seeds were ground to fineness for 5 min in a 240 W electrical blender (Panasonic MX-337, Malaysia) before extraction. A solvent-to-solid ratio of 10:1 was used in this study. Briefly, 50 g of dried ground *Carica papaya* cv. Sekaki seeds were weighed in a conical flask, and 500 mL of MeOH was added. The extraction was carried out at room temperature (27 $^\circ$C) for 8 h in a shaker (100 rpm) followed by filtration through Whatman No.1 filter paper (GE Healthcare, UK). The filtrate was transferred to pre-weighed flat bottom flasks and concentrated using a rotary vacuum evaporator (Eyela N-1001, Japan) at 40 $^\circ$C. The concentrated CPSE was kept at 4 °C until further use (Sani et al., 2017a).

8.2.3 Effect of Temperature on the Extract

About 29.5 mg of nitrogen-blown extract was heated at 60, 80, 100, 150, and 200 °C for 15 min and mixed with 1% Tween 80 and TSB for a final volume of 5 mL. The final extract concentration used in this study was 5.63 mg/mL, which was the MIC value determined by Sani et al. (2017b). Heated-extract solutions were subjected to the test of percentage growth inhibitions of *S. enteritidis, V. vulnificus, P. mirabilis,* and *B. cereus.*

8.2.4 Percentage of Growth Inhibition

The growth of *Salmonella enteritidis* (ATCC 13076), *Vibrio vulnificus* (ATCC 27562), *Proteus mirabilis* (ATCC 12453), and *Bacillus cereus* (ATCC 10875) was monitored using 96 well-microplates. A volume of 10 µL of TSB containing 10^6 CFU/mL of the tested pathogen was mixed with 190 µL of the TSB solutions in 96 well-microplates and assessed in a microplate spectrophotometer. Positive controls containing the tested solution was inoculated with the respective pathogens. Negative controls contained a mixture of crude extract, Tween 80, and TSB. The 96 well-microplate was incubated at 37 °C for 24 h on a Heidolph Inkubator and Titrama × 1000 (Germany) at 210 rpm to prevent adherence and clumping, after which the optical density was measured at 600 nm in Tecan Infinite® 200 Microplate Reader (Switzerland) before (T_0) and after (T_{24}) incubation.

The percentage growth inhibition (Patton et al., 2006) for TSB solutions was determined using Eq. (8.1)

$$\text{Growth Inhibition (PI)\%} = (1 - (\text{OD test well/OD of positive control well})) \times 100 \quad (8.1)$$

8.2.5   Effect of Temperature on Fatty Acid Profile in Carica Papaya Seed Extract

The CPSE was heated at 40, 60, 80, 100, 150, and 200 °C for 15 min and subjected to pretreatment before GC/MS analysis.

8.2.6   Profiling of Fatty Acid Methyl Esters by GC/MS Analysis

8.2.6.1   Sample Preparation

An amount of 0.01 g/mL of the heated extract at 40, 60, 80, 100, 150, and 200°C for 15 min was re-dissolved in 0.6 mL of hexane and added with 0.4 mL of 1 M sodium methoxide and vortexed for 30 s. The top hexane layer (0.6 mL) was subjected to FAMEs quantification by gas chromatography-mass spectrometry (GC/MS) analysis.

8.2.6.2   Preparation of Calibration Curve

The linearity of the methods was evaluated using different concentrations of FAME standards, ranging from 0.0005 – 3 mg/mL and cis-vaccenic acid (0.0001 – 0.5 mg/mL). Linearity was assessed using the linear regression equation, where the correlation coefficient r > 0.98, indicated an acceptable identification (Fagundes & Caldas, 2012). The prepared standards were analyzed using GC/MS.

8.2.6.3   Quantification of Fatty Acids Methyl Esters

Characterization of FAMEs was performed using the Agilent-Technologies 7890A GC system equipped with the Agilent-Technologies 5975 mass selective detector (Agilent Technologies, USA). The compound separation was achieved using an HP-88 capillary column (100 m × 0.25 mm, film thickness 0.20 μm) with an oven temperature program at 150 °C for 5 min, heated to 240 °C at the rate of 4 °C/min and held for 15 min. Samples were injected in split mode with the injector temperature at 260 °C with helium as a carrier gas at a constant flow rate (1 mL/min). For MS detection, the electron ionization mode with ionization energy of 70 eV was used with a mass range of m/z 20–700 units. The MS transfer line and MS quadrupole temperature were set at 230 °C and 150 °C. The mass spectrometer was operated in both the scan and selected ion monitoring (SIM) modes for compound identification and quantification, respectively. In order to avoid the need to modify the retention times in the calibration tables due to column maintenance or column change, the calibration of standards was performed in the retention-time-lock mode (Caven-Quantrill & Buglass, 2007) where palmitic acid (C16:0) was chosen as the locking standard due to its stability. Compounds were identified by their retention times and mass fragmentation patterns of standards using the National Institute of Standard (NIST) Mass Spectral 11 offline library.

8.2.7   Statistical Analysis

8.2.7.1   ANOVA

Data were expressed as mean $\pm$ standard deviations of triplicate R_f and LC_{50} and residual methanol. One way analysis of variance (ANOVA) with Tukey's test was conducted using XLSTAT-Pro (2014) statistical software (Addinsoft, Paris, France) to determine the significant difference between the means at 95% confidence level ($p < 0.05$) for bioautography, toxicity assay, and residual methanol.

8.2.7.2   Principle Component Analysis (PCA)

Principle component analysis (PCA) is the most commonly unsupervised pattern recognition technique used in the distribution of compounds in the food sample (Dorta et al., 2014). The PCA was employed to elucidate the data variance of inter-correlated variables and transform them into independent variables called a principle component (PC). PCA also excludes the less significant parameters.

In this study, PCA was applied to FAME's contribution to the PI of tested pathogens as affected by temperature. From the calculation of the eigenvalue, a new set of groups called PCs was generated for each eigenvalue > 1. The PC was influenced by factor loading > 0.75, 0.74 - 0.50, and 0.49 – 0.30, which were considered as strong, moderate, and weak (Retnam et al., 2013). The profile of factor loadings and specific indicative FAMEs were used to deduce the FAMEs contribution on PI of tested pathogens as affected by temperature. PCA was conducted using XLSTAT-Pro (2014) statistical software (Addinsoft, Paris, France).

8.3   Discussion

8.3.1   Effect of Temperature on Antibacterial Activity and Fatty Acids Profile of Carica Papaya Seed Extract

The effect of various heating temperatures on the antibacterial activity of CPSE indicated the stability of the extract as shown in Fig. 8.1. The potency of the extract

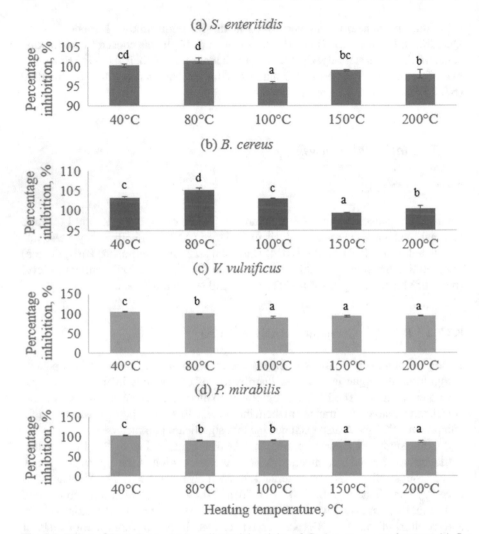

Fig. 8.1 Effect of heated extract on antibacterial activity of *Carica papaya* seed extract (a) *S. enteritidis*, (b) *B. cereus,* (c) *V. vulnificus* and (d) *P. mirabilis* growths

against *S. enteritidis* and *V. vulnificus* had the same characteristics where the extract heated > 100 °C had a percentage inhibition of < 100%. Only the heated extract at 150 °C indicated a percentage inhibition < 100% (99.45%) against *B. cereus*. For *P. mirabilis*, only the heated extract at 40 °C was potent to the pathogen because the percentage inhibition > 100%, even though Nychas et al. (2003) reported that low temperatures reduced antibacterial activities. However, He et al. (2010) found that most antibacterial agents had lost their inhibitory efficiency at high temperatures. In general, all tested pathogens were sensitive to the extract in TSB at < 40 °C and this finding could be proposed for food incorporated with the extract to be handled below this temperature before consumption.

8.3.2   Effect of Temperature on Fatty Acid Methyl Esters Profile of Carica Papaya Seed Extract

Linear relationships between the ratios of the peak area signals and the corresponding concentrations of FAMEs content were observed in the analytical curves Fig. 8.2 when using different concentrations. The parameters of the analytical curves with the correlation of determination (R^2) are shown in Table 8.1. The values of the analytical curve led to the conclusion that the linear regression model was adequate for the analytical determinations in this study, as R^2 was higher than 0.98 (Fagundes & Caldas, 2012). In this study, we did not calculate trans vaccenic acid, and thus only cis-vaccenic acid (C18:1n11c) was recorded. The chromatogram of the FAMEs is shown in Fig. 8.3.

The PCA was used to establish the relationship between the FAMEs identified by GC/MS and PI of *S. enteritidis*, *B. cereus*, *V. vulnificus,* and *P. mirabilis* as affected by heat. The four main principal components (PCs) characterized were having an eigenvalue > 1 (Saiful et al., 2019) which was considered as significant factor loadings (FL) ($p < 0.05$). The four PCs also had a cumulative explained total variance of 100% which consisted of PC1 (47.99%), PC2 (22.01%), PC3 (20.23%), and PC4 (9.77%) (Table 8.2). The data variances were explained at 70% for PC1 versus PC2 and 68.22% for PC1 versus PC3.

The FL table showed the loading values between the FAMEs and the PI of tested pathogens (Table 8.2). Based on the strong FL limit (> 0.75), PC1 was mainly characterized for a higher content of C16:0, C16:1, C18:0, C18:1n9c, C18:1n11c, and

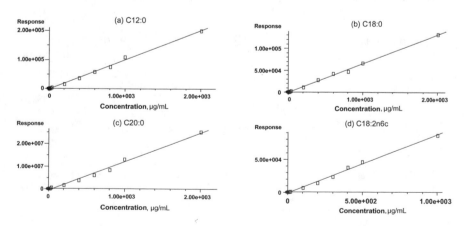

Fig. 8.2   Calibration curve of (a) C12:0, (b) C18:0, (c) C20:0 and (d) C18:2n6c

Table 8.1 Calibration information of fatty acid methyl esters

No	Compound[1]	Assignment	Mass	Rt^2	$(R^2)^3$	Linearity equation
1.	Butyric acid	C4:0	102	9.690	0.9974	$y = 0.3406x + 14.727$
2.	Hexanoic acid	C6:0	130	10.144	0.9966	$y = 0.0923x - 15.233$
3.	Octanoic acid	C8:0	158	10.940	0.9952	$y = 0.3662x + 59.885$
4.	Decanoic acid	C10:0	186	12.304	0.9969	$y = 2.0091x - 437.51$
5.	Undecanoic acid	C11:0	200	13.253	0.9945	$y = 1.5095x - 323.16$
6.	Dodecanic acid	C12:0	214	14.399	0.9966	$y = 4.0352x - 1227.9$
7.	Tridecanoic acid	C13:0	228	15.689	0.9962	$y = 1.481x + 26.513$
8.	Myristic acid	C14:0	242	17.148	0.9964	$y = 4.2624x - 1108$
9.	Myristoleic acid	C14:1	240	18.327	0.9962	$y = 0.4265x + 230.14$
10.	Pentadecanic acid	C15:0	256	18.670	0.9961	$y = 2.5957x - 136.37$
11.	Cis-10-pentadecenoic	C15:1	254	19.912	0.9946	$y = 0.4751x + 182.81$
12.	Palmitic acid	C16:0	270	20.285	0.9964	$y = 13.148x - 3445.9$
13.	Palmitoleic acid	C16:1	268	21.295	0.9885	$y = 0.4777x + 666.03$
14.	Heptadecanic acid	C17:0	284	21.822	0.9960	$y = 2.8421x + 105.16$

(continued)

Table 8.1 (continued)

No	Compound[1]	Assignment	Mass	Rt^2	$(R^2)^3$	Linearity equation
15.	Cis-10-heptadecenic acid	C17:1	282	22.877	0.9863	$y = 0.5708x + 420.79$
16.	Stearic acid	C18:0	299	23.418	0.9972	$y = 2.6297x - 466.87$
17.	Elaidic acid	C18:1n9t	296	24.007	0.9903	$y = 0.7292x + 523.17$
18.	Oleic acid	C18:1n9c	296	24.329	0.9908	$y = 1.3187x + 894.16$
19.	Linolelaidic acid	C18:2n6t	294	24.947	0.9916	$y = 1.7255x + 495.27$
20.	Linoleic acid	C18:2n6c	294	25.631	0.9947	$y = 1.7667x - 902.56$
21.	Arachidic acid	C20:0	327	26.430	0.9911	$y = 504.82x - 542642$
22.	γ-linolenic acid	C18:3n6	292	26.580	0.9923	$y = 155.49x + 166488$
23.	Linolenic acid	C18:3n3	292	27.142	0.9959	$y = 163.93x - 36714$
24.	Cis-11-eicosenoic acid	C20:1	325	27.259	0.9975	$y = 210.9x - 23947$
25.	Cis-vaccenic acid	C18:1n11c	282	27.613	0.9979	$y = 0.1722x + 7.134$
26.	Heneicosanoic acid	C21:0	341	27.835	0.9833	$y = 278.58x - 425782$
27.	Cis-11,14-eicosadienoic acid	C20:2	323	28.547	0.9950	$y = 177.5x - 51134$
28.	Behenic acid	C22:0	355	29.305	0.9904	$y = 670.73x - 984863$

(continued)

Table 8.1 (continued)

No	Compound[1]	Assignment	Mass	Rt[2]	(R^2)[3]	Linearity equation
29.	Cis-8,11,14-eicosatrienoic acid	C20:3n6	321	29.505	0.9973	y = 164.72x − 23659
30.	Cis-11,14,17-eicosatrienoic acid	C20:3n3	321	30.146	0.9918	y = 261.24x − 230243
31.	Erucic acid	C22:1n9	353	30.147	0.9965	y = 187.08x + 42244
32.	Arachidoic acid	C20:4n6	318	30.245	0.9926	y = 152.73x + 32586
33.	Tricosanic acid	C23:0	369	30.713	0.9900	y = 2.1268x − 1870.7
34.	Cis-13,16-docosadienoic acid	C22:2n6	351	31.517	0.9970	y = 1.2291x − 1040.4
35.	Cis-5,8,11,14,17-eicosapentaenoic acid (EPA)	C20:5n3	316	31.901	0.9917	y = 122.77x + 6381.4
36.	Tetracosanoic acid	C24:0	383	32.281	0.9842	y = 8.0799x − 11217
37.	Cis-15-tetracosenic acid	C24:1n9	381	33.234	0.9947	y = 232.97x − 73520
38.	Cis-4,7,10,13,16,19-docosahexaenoic acid (DHA)	C22:6n3	343	36.354	0.9908	y = 101.36x + 62011

[1] Fatty acids detected were in their esters (FAMEs) form
[2] Rt = retention time
[3] R^2 = coefficient determination

lower content of C23:0, also had strongly contributed to the PI of *V. vulnificus* and *P.mirabilis* and moderate contribution to *S. enteritidis*. The PC2 was related principally to the higher content of C13:0 and C20:2. PC3, on the other hand, was dominated by a higher content of C10:0 and a lower content of C20:0 and moderately contributed to the PI of *B. cereus*.

The observation plot allowed exploring the correlations between the PCs and the extract as affected by temperature (Fig. 8.4). The extract heated at 40 °C clearly showed a significant antagonistic correlation in PC1 against 100 °C, 150 °C, and 200

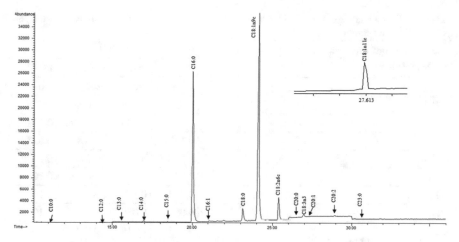

Fig. 8.3 Chromatogram of fatty acid methyl esters in *Carica papaya* seed extract separated by using HP88 column

°C, and significant antagonistic correlation against 80 °C in PC3 (Fig. 8.4b). In PC3 also, a significant antagonistic correlation was exhibited by 80 °C and 150 °C.

The variable plot in Fig. 8.5 helped to establish which FAMEs and PI of the tested pathogens discriminated against the heated extracts. The extract heated at 40 °C, as seen in Fig. 8.4a, had a strong PC1 score that has been related to higher content in C16:0, C16:1, C18:0, C18:1n9c, C18:1n11c (Fig. 8.5a). Nevertheless, the heated extract at 80 °C had a high PC3 score, as seen in Fig. 8.4b; therefore, had higher contents of C10:0 (Fig. 8.5b). The heated extract at 100 °C (Fig. 8.4a) had a higher content of C13:0 in Fig. 8.5a, whereas the extract heated at 150 °C had a lower content of C12:0 and C18:1n9t. However, the C23:0 was lower in 200 °C heated extracts.

These results suggest that the PI of the tested pathogens was addictively and antagonistically facilitated by different compositions of FAMEs. Besides, the individual FAMEs had different stability against different temperatures. In summary, from PC1 alone, we found that C16:0, C16:1, C18:0, C18:1n9c, and C18:1n11c from CPSE had strongly inhibited *V. vulnificus* and *P. mirabilis* growths and moderately inhibit *S. enteritidis* growth.

8.3.3   Profile of Cis and Trans Fatty Acid Methyl Esters as Affected by Temperature

FAMEs were dominant in CPSE (Sani et al., 2017b). Among the FAMEs, their trans form has received high concern due to its negative health impact, especially when the extract was used in the food and undergoes heat treatment, such as cooking and deep-frying. Thus, this study was done to identify the profile of the trans-FAMEs when

Table 8.2 Factor loadings of fatty acid methyl esters and percentage inhibition of tested pathogens as affected by temperature

Fatty acid methyl esters and percentage inhibition[1] of tested pathogens	Factor loadings (FL)[2,3,4]			
	PC1	PC2	PC3	PC4
C10:0	0.200	−0.253	**0.937**	−0.132
C12:0	−0.258	−0.733	−0.493	−0.391
C13:0	−0.611	**0.755**	−0.008	0.238
C14:0	0.137	−0.549	0.470	0.677
C15:0	−0.515	0.432	0.027	0.740
C16:0	**0.940**	−0.162	−0.274	0.122
C16:1	**0.898**	0.158	−0.383	0.148
C18:0	**0.989**	0.088	−0.068	0.101
C18:1n9c	**0.979**	0.016	0.026	0.201
C18:1n9t	−0.552	−0.667	−0.461	0.194
C18:2n6c	0.670	−0.620	0.043	0.406
C20:0	0.409	−0.092	**-0.890**	0.182
C18:3n3	−0.396	−0.720	0.404	−0.401
C20:1	−0.494	0.467	0.703	0.208
C18:1n11c	**0.877**	0.264	0.350	−0.194
C20:2	0.438	**0.776**	−0.359	−0.278
C23:0	**−0.952**	−0.011	−0.076	0.295
PI of *V. vulnificus*	**0.962**	−0.159	0.114	0.192
PI of *S. enteritidis*	*0.662*	−0.603	0.434	0.096
PI of *B. cereus*	0.539	0.375	*0.727*	−0.201
PI of *P. mirabilis*	**0.919**	0.356	−0.163	−0.051
Eigen value	10.078	4.623	4.248	2.051
Variability, %	47.99	22.01	20.23	9.77
Cumulative, %	47.99	70.01	90.23	100.00
Significance level, α	< 0.05	< 0.05	< 0.05	< 0.05

[1] PI = percentage inhibition
[2] The FL were considered strong (> 0.75), moderate (0.74 - 0.50) and weak (0.49 − 0.30)
[3] Strong FL correlation > 0.75 was shown in bold
[4] Moderate FL correlation (0.74 - 0.50) for PI of *S. enteritidis* and *B. cereus* was shown in italic

the extract is heated. Also, the study of other FAMEs profile in non-heated CPSE (unheated) *Carica papaya* seed was done due to the capability of the HP-88 column to separate cis and trans-FAMEs, unlike the HP-5 ms column used in common plant metabolites analysis using GC/MS.

The profile of cis and trans-FAMEs affected by temperature is shown in Fig. 8.6. The highest concentration of oleic acid (C18:1n9c) was recorded at low temperatures

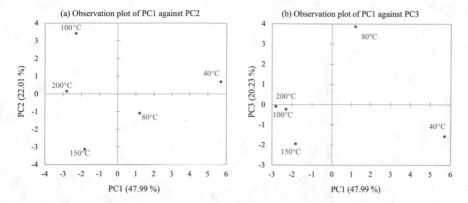

Fig. 8.4 Observation plot of (a) PC1 and PC2 and (b) PC1 and PC3 of extract distribution as affected by temperature

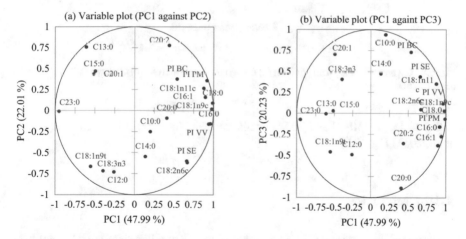

Fig. 8.5 Variable plot resulting from (a) PC1 against PC2 and (b) PC1 against PC3 of fatty acid methyl esters and PI of tested pathogens present in extract as affected by temperature

(40 °C and 60 °C), whereas cis-vaccenic acid (C18:1n11c) was detected at each heating treatment and exhibited a reducing trend; C18:1n11c was the most stable FAMEs at 100 °C whereas other FAMEs were undetected.

C18:2n6t was not detected in each sample treatment, indicating that C18:2n6c was stable against heat and did not convert to trans FAMEs. Meanwhile, C18:1n9c showed a drastic reduction when heated at higher temperatures and producing its trans form (C18:1n9t) at 150 °C and 200 °C, thereby supports the finding of Gürdeniz et al. (2013) since all naturally occurring FAMEs of plant origin are in the cis form, and the trans form is generally generated when oils and fats are hydrogenated or heated at a high temperature (Tao, 2007). Thus, it can be proposed that the food incorporated with the CPSE could be handled at a temperature < 150 °C. However, C18:1n9c was

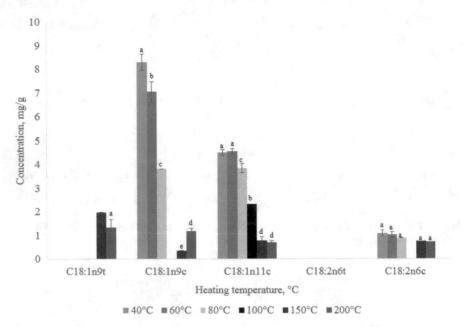

Fig. 8.6 Cis and trans fatty acids profile; (a) C18:1n9t, (b) C18:1n9c, (c) C18:1n11c, (d) C18:2n6t, (d) C18:2n6c

still detected at 150 °C and 200 °C because of its high oxidative stability (Preedy et al., 2011) Fig. (8.6).

8.4 Conclusion

In summary, the crude of CPSE cv. Sekaki/Hong Kong had demonstrated antibacterial activity against *S. enteritidis, V. vulnificus, P. mirabilis*, and *B. cereus*. These tested pathogens were sensitive to the extract in TSB at < 40 °C. From the PCA, C16:0, C16:1, C18:0, C18:1n9c, and C18:1n11c from CPSE had strongly inhibited *V. vulnificus* and *P. mirabilis* growths and moderately inhibit *S. enteritidis* growth. The treatment of CPSE against heat had also caused the generation of the trans-FAMEs at 150 °C and 200 °C, thus indicating that CPSE should be handled at < 150 °C in food applications.

Acknowledgements This work was supported by the Fundamental Research Grant Scheme (FRGS19-041-0649) of the Ministry of Higher Education Malaysia and Research University Grant (vot no. 9328200) of Universiti Putra Malaysia.

References

Albuquerque, T. G., Costa, H. S., Castilho, M. C., & Sanches-Silva, A. (2011). Trends in the analytical methods for the determination of trans fatty acids content in foods. *Trends in Food Science & Technology, 22,* 543–560. https://doi.org/10.1016/j.tifs.2011.03.009.

Caven-Quantrill, D. J., & Buglass, A. J. (2007). Determination of volatile organic compounds in English vineyard grape juices by immersion stir bar sorptive extraction gas chromatography-mass spectrometry. *Flavour and Fragrance Journal, 22*(November), 206–213. https://doi.org/10.100 2/ffj.

Chávez-Quintal, P., González-Flores, T., Rodríguez-Buenfil, I., & Gallegos-Tintoré, S. (2011). Antifungal activity in ethanolic extracts of Carica papaya L. cv. Maradol leaves and seeds. *Indian Journal of Microbiology, 51*(1), 54–60. https://doi.org/10.1007/s12088-011-0086-5.

Dorta, E., González, M., Lobo, M. G., Sánchez-Moreno, C., & de Ancos, B. (2014). Screening of phenolic compounds in by-product extracts from mangoes (Mangifera indica L.) by HPLC-ESI-QTOF-MS and multivariate analysis for use as a food ingredient. *Food Research International, 57,* 51–60. https://doi.org/10.1016/j.foodres.2014.01.012.

Fagundes, A. M., & Caldas, S. S. (2012). Development and validation of a method for the determination of fatty acid methyl ester contents in tung biodiesel and blends. *Journal of the American Oil Chemists' Society, 89,* 631–637. https://doi.org/10.1007/s11746-011-1948-z.

FAOSTAT. (2019). *Papaya crop yield.* Retrieved May 30, 2019, from http://www.fao.org/faostat/en/#data/QC.

Gürdeniz, G., Rago, D., Bendsen, N. T., Savorani, F., Astrup, A., & Dragsted, L. O. (2013). Effect of trans fatty acid intake on LC-MS and NMR plasma profiles *PLoS ONE, 8*(7), 1–11. https://doi.org/10.1371/journal.pone.0069589.

He, F., Yang, Y., Yang, G., & Yu, L. (2010). Studies on antibacterial activity and antibacterial mechanism of a novel polysaccharide from Streptomyces virginia H03. *Food Control, 21*(9), 1257–1262. https://doi.org/10.1016/j.foodcont.2010.02.013.

Lim, T. K. (2012). Carica papaya. In *Edible Medicinal and Non-Medicinal Plants* (pp. 693–717). https://doi.org/10.1007/978-90-481-8661-7.

Nakamura, Y., Yoshimoto, M., Murata, Y., Shimoishi, Y., Asai, Y., Eun, Y. P., ... Nakamura, Y. (2007). Papaya seed represents a rich source of biologically active isothiocyanate. *Journal of Agricultural and Food Chemistry, 55*(11), 4407–4413. https://doi.org/10.1021/jf070159w.

Nychas, G. J. E., Skandamis, P. N., & Tassou, C. C. (2003). Antimicrobials from herbs and spices. In S. Roller (Ed.), *Natural antimicrobials for the minimal processing of foods* (1st ed.). Woodhead Publishing Limited. https://doi.org/10.1016/B978-1-85573-669-6.50014-9.

Patton, T., Barrett, J., Brennan, J., & Moran, N. (2006). Use of a spectrophotometric bioassay for determination of microbial sensitivity to manuka honey. *Journal of Microbiological Methods, 64*(1), 84–95. https://doi.org/10.1016/j.mimet.2005.04.007.

Preedy, V. R., Watson, R. R., & Patel, V. B. (2011). *Nuts and seeds in health and disease prevention.* In V. R. Preedy, R. R. Watson, & V. B. Patel (Eds.). (1st ed.). Elsevier.

Retnam, A., Zakaria, M. P., Juahir, H., Aris, A. Z., Zali, M. A., & Kasim, M. F. (2013). Chemometric techniques in distribution, characterisation and source apportionment of polycyclic aromatic hydrocarbons (PAHS) in aquaculture sediments in Malaysia. *Marine Pollution Bulletin, 69*(1–2), 55–66. https://doi.org/10.1016/j.marpolbul.2013.01.009.

Saiful, M., Azid, A., Iskandar, S., Shirwan, M., Sani, A., & Lananan, F. (2019). Comparison of prediction model using spatial discriminant analysis for marine water quality index in mangrove estuarine zones. *Marine Pollution Bulletin, 141*(February 2018), 472–481. https://doi.org/10.1016/j.marpolbul.2019.02.045.

Sani, M. S. A. (2018). *Antibacterial activities of Carica papaya L. seed as food preservative.* Universiti Putra Malaysia. Universiti Putra Malaysia. Retrieved from http://www.elib.upm.edu.my/cgi-bin/koha/opac-ISBDdetail.pl?biblionumber=574282.

Sani, M. S. A., Bakar, J., Rahman, R. A., & Abas, F. (2017a). In vitro antibacterial activities and composition of Carica papaya cv. Sekaki / Hong Kong peel extracts. *International Food Research Journal, 24*(June), 976–984.

Sani, M. S. A., Bakar, J., Rahman, R. A., & Abas, F. (2017b). The antibacterial activities and chemical composition of extracts from Carica papaya cv. Sekaki / Hong Kong seed. *International Food Research Journal, 24*(April), 810–818.

Tao, B. Y. (2007). Industrial applications for plant oils and lipids. In S.-T. Yang (Ed.), *Bioprocessing for value-added products from renewable resources* (1st ed., pp. 611–627). San Diego: Elsevier B.V. http://dx.doi.org/10.1016/B978-044452114-9/50025-6.

Vasu, P., Savary, B. J., & Cameron, R. G. (2012). Purification and characterization of a papaya (Carica papaya L.) pectin methylesterase isolated from a commercial papain preparation. *Food Chemistry, 133*(2), 366–372. https://doi.org/10.1016/j.foodchem.2012.01.042.

Chapter 9
Solid-State Fermentation of Agro-Industrial Waste Using Heterofermentative Lactic Acid Bacteria

Lina Oktaviani, Muhammad Yusuf Abduh, Dea Indriani Astuti, and Mia Rosmiati

Abstract Solid-state fermentation is commonly applied to valorize agro-industrial waste to produce various bioproducts. Good solid-state fermentation highly depends on the development of inoculum before the fermentation. The development of inoculum using suitable bacteria and the appropriate age of inoculum may reduce the lag phase and accelerate bacterial growth during fermentation. This chapter describes the solid-state fermentation of agro-industrial wastes using heterofermentative lactic acid bacteria with a focus on the stage of inoculum development, including preparation of medium, activation of bacteria, and growth curve analysis, to determine bacterial growth rate during the fermentation. This process can also be applied to other complex solid substrates.

Keywords Agro-industrial waste · Bacterial growth · Heterofermentative lactic acid bacteria · Inoculum · Solid-state fermentation

Abbreviation

HCl	Hydrogen chloride
MRS	Man, Rogosa, and Sharpe
MRSB	Man, Rogosa, and Sharpe broth
MRSB-S	Man, Rogosa, and Sharpe broth supplemented with the solid substrate
NaOH	Sodium hydroxide
SSF	Solid-state fermentation

L. Oktaviani · M. Y. Abduh (✉) · D. I. Astuti · M. Rosmiati
Institut Teknologi Bandung, 40132 Bandung, Indonesia
e-mail: yusuf@sith.itb.ac.id

© The Author(s), under exclusive license to Springer Nature Switzerland AG 2021
A. Amid (ed.), *Multifaceted Protocols in Biotechnology, Volume 2*,
https://doi.org/10.1007/978-3-030-75579-9_9

9.1   Introduction

Solid-state fermentation (SSF) is a process of microbial growth on solid materials in the near absence of free water. SSF is a widely used technique to valorize agro-industrial waste to produce various bioproducts (Singhania et al., 2009; Oktaviani et al., 2019). SSF using agro-industrial waste is inherently difficult due to lack of nutrients and antimicrobial compounds that result in cell feedback inhibition, as is often the case with bacteria as fermentation agents (Lizardi-Jimenez & Hernandez-Martinez, 2017). The development of the inoculum has become an important stage to support bacterial growth during SSF. The development of the inoculum ensures that bacteria grow at a faster rate and high biomass concentrations are achieved at the beginning of fermentation (Nyalang, 2005).

The development of inoculum should emphasize methods aimed at minimizing the lag phase and increase the metabolically active cells (Nyalang, 2005). This can be achieved using highly adaptive bacteria that have already entered the final exponential phase (Maier & Pepper, 2015). This chapter will discuss the SSF procedure for the use of bacteria from the preparation to the fermentation process with a focus on the inoculum development stage.

The preparation stage consists of medium preparation and inoculum preparation. Medium preparation includes making a general medium and an adapted medium. The preparation of inoculum includes bacterial activation and the development of inoculum focuses to achieve ideal bacterial conditions using the growth curve approach. In the main fermentation process, bacterial growth is monitored and compared to the use of bacteria without proper inoculum preparation. In this procedure, the fermentation substrate was coffee pulp as an agro-industrial waste whereas the bacteria used were heterofermentative lactic acid bacteria from *Leuconostoc sp.*

9.2   Objective

This chapter investigates the SSF of coffee pulp using heterofermentative lactic acid bacteria from *Leuconostoc sp.* The discussion concentrates on the stages of inoculum development and its effects on bacterial growth during the SSF.

9.3   Methodology

9.3.1   Preparation of Medium for Activation of Bacteria

The survival and continued growth of microorganisms depend on an adequate supply of medium. Bacteria stored in a preservation agar medium need to be activated into a medium broth. To support the initial growth of bacteria, the medium used at the beginning of bacterial activation should be a rich medium, such as Man, Rogosa, and Sharpe (MRS) and M17 for lactic acid bacteria, or it can be a general medium such as nutrient broth (NB) (Cappuccino & Sherman, 2014). A specific medium such as MRS medium exhibit consistent growth of lactic acid bacteria, especially for *Lactobacilli*. During medium preparation, an antifungal solution can be added to reduce the risk of contamination.

9.3.2   Bacterial Growth Curve

Bacterial inoculum requires an adapted condition and an optimum age before SSF begins. These conditions can reduce the potential environmental shock due to the transfer of bacteria from a rich agar medium to an SSF medium. The medium used during the inoculum development must be adapted to the compounds contained in the fermentation substrate (Roszak & Colwell, 1987). The use of this medium can support bacterial cells to adapt and grow rapidly under stress conditions during SSF (Roszak & Colwell, 1987).

The optimum inoculum age can be reached at the maximum growth rate of bacteria. It can be analyzed through an exponential phase or logarithmic phase in the bacterial growth curve (Fig. 9.1). In the exponential phase, the bacteria will

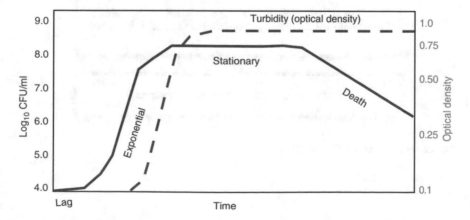

Fig. 9.1 A typical growth curve for bacteria (Maier & Pepper, 2015)

grow rapidly under the conditions present in the system and the increase in bacterial biomass will be proportional to the initial biomass concentration (Maier & Pepper, 2015). The bacterial growth rate can be calculated using Eq. (9.1) and the bacterial growth phase can be determined using a bacterial growth curve.

$$v = \frac{C_E - C_0}{t_E - t_0} \tag{9.1}$$

Where v is the bacterial growth rate in CFU ml^{-1} h^{-1}, C is the number of bacterial cells in cells unit, and t is time in hour unit, C_o is the bacterial cell at time zero (t_o) and C_E is the number of bacterial cells at end of specific time (t_E).

Optimum inoculum age can be obtained from the exponential phase. This phase occurs after the lag phase and before the stationary phase. The increase in numbers or bacterial mass can be measured as a function of time to obtain a growth curve. Several distinct growth phases can be observed within a growth curve (Fig. 9.2). These include the lag phase, the exponential or log phase, the stationary phase, and the death phase. Each of these phases represents a distinct period of growth associated with typical physiological changes in the cell culture. As can be seen in the following sections, the growth rates associated with each phase are quite different.

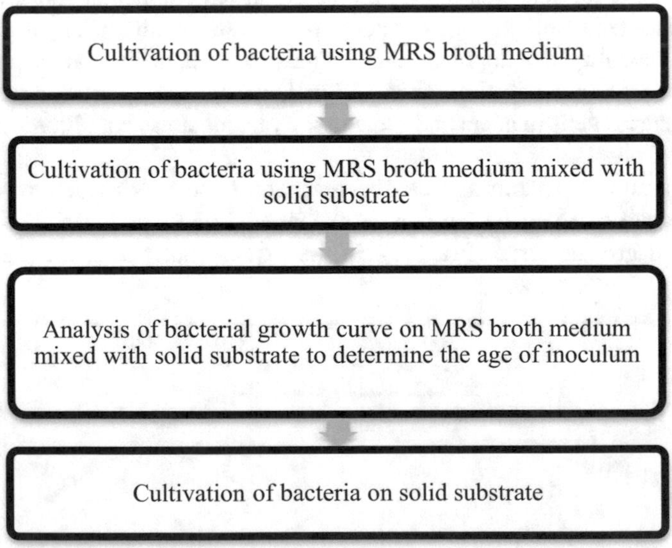

Fig. 9.2 Flowchart for solid-state fermentation

9.3.3 Heterofermentative Lactic Acid Bacteria

Lactic acid bacteria can be found in any material rich in carbohydrates such as food waste and agro-industrial waste which provide a potential source of nutrients for growth of lactic acid bacteria and production of valuable compounds. There are 2 types of lactic acid bacteria based on the fermentation pathway, homofermentative and heterofermentative. Homofermentative lactic acid bacteria produce lactic acid as the main product while heterofermentative lactic acid bacteria can produce varied compounds such as lactic acid, ethanol, acetic acid, and other valuable compounds. These varied products are an advantage for the utilization of heterofermentative lactic acid bacteria. The bacteria are facultative anaerobic bacteria that can grow in aerobic and anaerobic conditions which also give a flexibility of the fermentation conditions (Victoria et al., 2012).

9.4 Materials

Consumable Items
Man, Rogosa, and Sharpe broth medium from HiMedia
Distilled water

Equipment
Erlenmeyer flask
Tea strainer
Magnetic stirrer
Perforated plate

9.5 Methods

The general steps for fermentation are shown in Fig. 9.2 whereas all detailed steps are explained in methods 9.5.1 to 9.5.5.

9.5.1 Preparation of MRS Broth Medium (MRSB)

1. Suspend 55.15 g of MRSB powder in 1000 mL of distilled water.
2. Heat the medium and stir vigorously using a magnetic stirrer to dissolve the medium completely.
3. Pour the medium into an Erlenmeyer flask. The maximum volume of the medium is 30% of the Erlenmeyer flask volume.
4. Sterilize the medium in an autoclave machine at 121 °C and 15 psi for 15 min.

5. Add antifungal solution such as nystatin on the sterile medium aseptically. The maximum concentration of the antifungal solution is 50 u/L of the medium.

9.5.2 Preparation of MRS Broth Medium Mixed with Solid Substrate (MRSB-S)

1. Suspend 55.15 g of MRSB powder in 1000 mL of distilled water.
2. Heat the medium and stir vigorously using a magnetic stirrer to dissolve the medium completely.
3. Add 1% (w/v) dry weight basis of the solid substrate into the MRSB medium, stir vigorously during the heating process for 30 min. The solid substrate in this procedure is fresh coffee pulp. The percentage of the solid substrate in the broth medium can be varied depending on the material used. For materials with high phenolic compounds such as coffee pulp, a percentage of 1% (w/v) is sufficient. Percentages above 1% (w/v) produce a medium with a higher phenolic compounds concentration which inhibits the growth of bacteria.
4. Filter the medium using a tea strainer before the medium is poured into an Erlenmeyer flask.
5. Wait until the temperature of the medium is below 40 °C. Check the pH of the medium, adjust the pH until its value is the same as the pH of the solid substrate. In this procedure, the pH of coffee pulp is about 4.5–5.
6. Pour the medium into an Erlenmeyer flask. The maximum volume of the medium is 30% of the volume of the Erlenmeyer flask.
7. Sterilize the medium using an autoclave machine at 121 °C and 15 psi for 15 min.

9.5.3 Activation of Bacteria and Growth Curve Analysis

1. Swab bacterial culture (*Leuconostoc sp.*) with a sterile loose loop, place one to two loopful of the cell suspension to an Erlenmeyer flask containing 100 mL MRSB medium.
2. Incubate the culture in an incubator shaker set at 37 °C and 130 rpm for 24 h.
3. Inoculate the culture from step 2 into a sterile MRSB medium. Inoculum size is 10% v/v with concentration of 10^6–10^7 CFU/mL. Incubate the culture at 37 °C and 130 rpm for 24 h.
4. Inoculate the culture from step 3 into a sterile MRSB-S medium. Inoculum size is 10% (v/v) with concentration of 10^6–10^7 CFU/mL. Incubate the culture at 37 °C and 130 rpm for 24 h.
5. The inoculum from step 4 is cultivated with a new and sterile MRSB-S medium. The inoculum size and incubation conditions are the same as in step 4.
6. Take 1 mL of the sample every 12 h to count the amount of bacterial cell which will be used for growth curve analysis.

7. The measurement of bacterial concentration in this medium uses a colony-forming unit. This method is used to quantify only the living bacteria. The sample is cultivated into an MRS agar medium aseptically using serial dilution and standard plate count method. The number of the bacterial colony is measured after incubation at 37 °C for 48 h.
8. Draw a growth curve with the y-axis is the number of bacterial cells (the log CFU/mL unit) and the x-axis is incubation time (h).
9. Determine the growth rate of bacteria using Eq. (9.1), inoculum age can be determined from the highest growth rate of bacteria.

9.5.4   Solid-State Fermentation Using Bacterial Inoculum

1. Inoculate bacterial inoculums obtained in the previous steps in a sterile container containing solid substrate (coffee pulp) and mixed well until the surface of coffee pulp is covered with bacterial inoculum. The size of the inoculum is 10% (v/w).
2. Spread the solid substrate onto a perforated plate to maintain aerobic condition. Make sure that the solid substrate is spread to form a layer of 1 cm in height.
3. Incubate the culture at 30–37 °C for a minimum of 4 h.
4. Collect 1 g of the sample after the incubation to measure the number of bacteria.

9.5.5   Measurement of Bacterial Concentration Using Serial Dillution and Standard Plate Count Method

1. Dilute 1 g of sample in 9 mL of a sterile physiological solution that contains 0.85% NaCl.
2. Make a dilution series from the sample.
3. Pipette out 0.1 mL from the appropriate dilution series onto the center of the agar plate surface
4. Dip an L-shaped glass spreader into alcohol and flame the glass spreader.
5. Spread the sample over the surface of the agar plate using the sterile glass spreader.
6. Incubate the plate at 37 °C for 48 h.
7. Calculate the amount of the bacterial colonies formed in the agar plate. Multiply the number of colonies by an appropriate dilution factor to determine the number of bacteria/mL in the original sample.

9.6 Results and Discussion

9.6.1 Activation of Bacteria from an Agar Media to a Broth Media

Bacterial transfer from the agar to the broth medium is the beginning of an activation stage, which is also called bacterial rejuvenation. When bacteria were still cultivated in the preservation of the agar medium, the functional abilities of the bacteria decline with time (Proenca et al., 2018). Therefore, the use of bacteria directly from the agar medium without repeated cultivation in a broth medium is not recommended. As evidence, Fig. 9.3 shows the growth of *Leuconostoc sp.* in the medium after being transferred from the agar medium. From Fig. 9.3, it can be seen that the lag phase of the bacteria is quite long. The number of bacteria increased after 18 h of incubation and the exponential phase began in the incubation period of 20–24 h with an average growth rate of 2.3×10^7 CFU ml^{-1} h^{-1}.

This lag phase is relatively longer than what has been reported in a similar study that *Leuconostoc* sp. cells would increase after 6 h of incubation in MRSB medium (Hamasaki et al., 2003). The long lag phase, as shown in Fig. 9.3, may be caused by bacterial aging due to the starvation of the bacteria. The problem arises because there is a limitation of growth on agar plates as solid substrate compared to liquid substrate. The growth limitation caused an abnormal reproduction of cells. Furthermore, the bacterial cells grew asymmetry between old and new cells, which results in different doubling times between the cells and consequently decreases the growth rate (Proenca

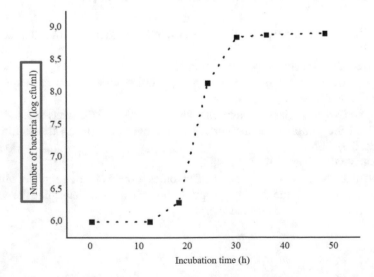

Fig. 9.3 Growth curve of *Leuconostoc sp.* cultivated using MRS broth medium in the initial activation stage

et al., 2018). To avoid this problem, bacterial cultures from agar to broth medium need to be re-grown in a new broth medium.

9.6.2 Growth Curve of Bacteria Cultivated Using MRS Broth Mixed with Coffee Pulp

Figure 9.4 shows the culture of *Leuconostoc sp.* in the MRSB-S medium. Initially, the MRSB-S medium is translucent dark brown, but after fermentation, the medium became turbid due to the growth of bacteria. The growth curve of *Leuconostoc sp.* cultivated using MRSB-S medium is shown in Fig. 9.5 which highlights that the lag phase was barely observed in this study. The lag phase is the stage where

Fig. 9.4 Culture of *Leuconostoc sp.* in MRS broth medium mixed with coffee pulp

Fig. 9.5 Growth curve of *Leuconostoc sp.* on MRS broth medium mixed with coffee pulp

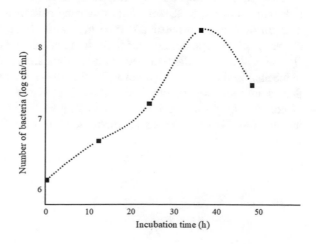

Table 9.1 Effect of different inoculum media to bacterial growth on solid substrate

Medium of Inoculum	The number of bacteria (log CFU/mL)	
	Initial (0 h of fermentation)	Final (4 h of fermentation)
MRS broth	5,97 ± 0,09	5,81 ± 0,13
MRS broth mixed with solid substrate	5,99 ± 0,20	8,18 ± 0,11

bacteria adjust their physiological and regulatory processes in order to adapt to a new environment. A short lag phase indicates that the bacteria have adapted well to the new environment (Schultz & Kishony, 2013). In this procedure, the new environment is the change of nutrients in the MRSB medium and pH after the addition of coffee pulp. MRSB-S medium has a low pH and rich with water-soluble components derived from the coffee pulp such as tannin, pectin, caffeine, and phenolic acid. Repeated cultivation procedures explained in Sect. 9.5.3 points 4 and 5 allow the bacteria to adapt to a new environment rapidly and consequently, the bacteria grew without a noticeable lag phase.

9.6.3 Growth of Bacteria During Solid-State Fermentation

During the fermentation process, the bacterial concentration was measured at the beginning and the end of the fermentation (4 h). The growth of bacterial inoculum from the MRSB-S medium is compared to the growth of bacterial inoculum from the MRSB medium (Table 9.1). The growth of the bacteria in the MRSB-S medium shows an increase in bacterial concentration relative to the growth of the bacteria in the MRSB medium.

The relatively stagnant concentration of bacteria during fermentation using inoculum from the MRSB medium might be caused by an environmental shock. The drastic change in nutrient composition and pH of the substrate decreases the efficiency of bacteria during acclimatization and adaptation to the new environment (Jaapar et al., 2011). Bacteria need time to adapt and produce the required enzymes to metabolize the substrate (Vogel & Todaro, 1984). Bacteria cultivated with MRSB-S medium increased rapidly because the bacteria adapted well to the conditions of coffee pulp during the inoculum preparation. Besides, the optimum age of the inoculum in this study boosts the multiplication of bacterial cells (Hornbæk et al., 2004).

9.7 Conclusions

Inoculum preparation plays an important role in bacterial growth during SSF. The use of a medium mixed with a solid substrate during the inoculum development stage increased the ability of the bacteria to grow rapidly during the fermentation. The concentration of bacteria can be measured using the standard plate count method and the growth rate can be determined with the help of a bacterial growth curve. SSF using different medium may be applied but with a prolonged observation period for more accurate results.

Acknowledgements The authors would like to acknowledge the Ministry of Research and Technology and Higher Education, the Republic of Indonesia for funding this study (002/SP2H/PTNBH/DRPM/2019).

References

Cappuccino, J. G., & Sherman, N. (2014). *Microbiology: Laboratory manual*. Pearson Education Inc.

Hamasaki, Y., Ayaki, M., Fuchu, H., Sugiyama, M., & Morita, H. (2003). Behavior of psychrotrophic lactic acid bacteria isolated from spoiling cooked meat products. *Applied and Environmental Microbiology, 69*(6), 3668–3671. https://doi.org/10.1128/AEM.69.6.3668-3671.2003.

Hornbæk, T., Nielsen, A. K., Dynesen, J., & Jakobsen, M. (2004). The effect of inoculum age and solid versus liquid propagation on inoculum quality of an industrial Bacillus licheniformis strain. *FEMS Microbiology Letters, 236*(1), 145–151. https://doi.org/10.1016/j.femsle.2004.05.035.

Jaapar, S. Z. S., Kalil, M. S., Ali, E., & Anuar, N. (2011). Effects of Age of Inoculum, Size of Inoculum and Headspace on Hydrogen Production using Rhodobacter sphaeroides. *Bacteriology Journal, 1,* 16–23. https://doi.org/10.3923/bj.2011.16.23.

Lizardi-Jimenez, M. A., & Hernandez-Martinez, R. (2017). Solid state fermentation (SSF): diversity of applications to valorize waste and biomass. *3 Biotech, 7*(44), 1–9. https://doi.org/10.1007/s13 205-017-0692-y.

Maier, R. M., & Pepper, I. L. (2015). Bacterial Growth. In *Environmental microbiology: Third edition* (pp. 37–56). https://doi.org/10.1016/B978-0-12-394626-3.00003-X.

Nyalang, C. E. (2005). *Inoculum development of bacteria for the decomposition of domestic waste.*

Oktaviani, L., Taufik, I., & Abduh, M. Y. (2019). Production of concentrated ruminant feed from cofee pulp and rice bran fermented by white oyster mushroom (*Pleurotus ostreatus*). *Jurnal Mikologi Indonesia, 13*(1), 15–29. https://www.jmi.mikoina.or.id/jmi/article/view/50.

Proenca, A. M., Rang, C. U., Buetz, C., Shi, C., & Chao, L. (2018). Age structure landscapes emerge from the equilibrium between aging and rejuvenation in bacterial populations. *Nature Communications, 9*(1). https://doi.org/10.1038/s41467-018-06154-9.

Roszak, D. B., & Colwell, R. R. (1987). Survival strategies of bacteria in the natural environment. *Microbiological Reviews, 51*(3), 365–379. https://doi.org/10.1002/sres.3850040205.

Schultz, D., & Kishony, R. (2013). Optimization and control in bacterial Lag phase. *BMC Biology, 11*(120), 1–3. https://doi.org/10.1186/1741-7007-11-120.

Singhania, R. R., Patel, A. K., Soccol, C. R., & Pandey, A. (2009). Recent advances in solid-state fermentation. *Biochemical Engineering Journal, 44*(1), 13–18. https://doi.org/10.1016/j.bej.2008.10.019.

Victoria, M., Valentin, L., & Renaulta, P. (2012). Genome sequence of Leuconostoc pseudome-
    senteroides strain 4882, isolated from a dairy starter culture. *Journal of Bacteriology, 194*(23),
    6637–6637. https://doi.org/10.1128/JB.01696-12.
Vogel, H. C., & Todaro, C. M. (1984). Fermentation and biochemical engineering handbook. In
    Elsevier Inc. (Vol. 28). https://doi.org/10.1016/0300-9467(84)85064-9.

Chapter 10
Synthesis of Chitosan-Folic Acid Nanoparticles as a Drug Delivery System for Propolis Compounds

Marselina Irasonia Tan and Adelina Khristiani Rahayu

Abstract This chapter discusses the method to synthesize chitosan-folic acid nanoparticles. Developments in nanotechnology provide an alternative drug delivery system. Chitosan is a polymer that can be utilized to make nanoparticles because it is biodegradable, non-toxic and inexpensive. These characteristics are important for the drug delivery system. To optimize this system, a specific ligand conjugated to the nanoparticle can aid in directing the nanoparticle to the cell target. Folic acid can be used as a ligand to direct nanoparticles to cell targets with high folic acid receptors such as tumor cells. In this research, a chitosan nanoparticle was synthesized to deliver propolis to the cell target. Since propolis bioavailability in the body is relatively low, its bioavailability needs to be improved by encapsulating it in nanoparticles. The purpose of this study is to synthesize folic acid conjugated chitosan nanoparticles that encapsulate propolis compounds. The effect of the molar ratio between chitosan:folic acid:sodium tripolyphosphate (TPP), chitosan molecular weight, encapsulated propolis concentration, sonication in the synthesis of chitosan-folic acid-containing propolis nanoparticles (NP-KF-P), and chitosan-folic acid conjugate nanoparticles (NP-KF-blanks) were also studied. The encapsulation efficiency of propolis in NP-KF and Fourier Transform Infrared Spectrophotometer (FTIR) of the nanoparticles was also observed. NP-KF-P and NP-KF-blanks were successfully synthesized by the ionic gelation method. The diameters of NP-KF-P and NP-KF-blanks were 153.9 ± 1.3 and 129.0 ± 3.4 nm, respectively. Propolis encapsulation efficiency in NP-KF-P was 30.37–73.36%, and 90% of the propolis could be released at pH 4.

Keywords Nanoparticles · Chitosan · Folic acid · Propolis

M. I. Tan (✉) · A. K. Rahayu
Institut Teknologi Bandung, Bandung, Indonesia
e-mail: marsel@sith.itb.ac.id

© The Author(s), under exclusive license to Springer Nature Switzerland AG 2021
A. Amid (ed.), *Multifaceted Protocols in Biotechnology, Volume 2*,
https://doi.org/10.1007/978-3-030-75579-9_10

10.1   Introduction

Nanoparticle technology had recently become an innovation in the drug delivery system development. Nanoparticles as drug delivery systems assure that drugs encapsulated in nanoparticles can reach the correct target cells (Vllasaliu et al., 2013). Nanoparticles can be targeted to a particular tissue/cell by a passive or active targeting system. In an active drug delivery system, the nanoparticle is conjugated with a specific ligand, which can help the nanoparticle to recognize the targeted receptor in the target area. In passive drug delivery systems, the nanoparticle reaches its destination by utilizing the target tissue's physiological condition, without any specific ligand (Vllasaliu et al., 2013). Moreover, it is possible to adjust the nanoparticle size, as well as the surface properties of nanoparticles and the release of active substances from nanoparticles. Due to their diminutive size, nanoparticles could easily pass through blood vessels and enter target cells (Mohanraj & Chen, 2006). Therefore, the use of nanoparticles can maximize the effect of drugs on target cells and minimize drug accumulation in healthy tissues.

10.2   Principle

Nanoparticles can be synthesized from various components, including lipids, polymers, and inorganic substances (Zhang et al., 2006). For biomedical purposes, nanoparticles must be biocompatible and non-toxic (Wilczewska et al., 2012). Chitosan is one of the components that can be used as nanoparticles. Chitosan has a low level of cytotoxicity and is degradable (Kean & Thanou, 2010). Therefore, chitosan is very suitable to use as a nanoparticle component for biomedical purposes. There are various kinds of chitosan with different molecular weights and degrees of deacetylation. Both of these features affect the biodegradable properties of chitosan (Kean & Thanou, 2010).

Chitosan (β(1–4) 2-amino 2-deoxy β-d glucan) is a polysaccharide obtained by chitin deacetylation (Fig. 10.1; Schmitz et al., 2019). Chitosan is composed of b(1–4)-linked D-glucosamine monomer and N-acetyl-D-glucosamine monomer (Riva et al., 2011). Chitin and chitosan support components of various organisms such as the exoskeletons of crustaceans and insects and parts of the fungal cell wall (Riva et al., 2011).

Chitosan is soluble in an acidic solution (pH below 6.5) due to its amine groups becoming protonated (Riva et al., 2011). Chitosan solubility in acidic conditions is beneficial in drug delivery systems, because these characteristics can be applied to regulate drug release from chitosan nanoparticles (Cheng et al., 2017; Viviek et al., 2013; Wang et al., 2017).

Chitosan nanoparticles can be formed with the help of tripolyphosphate (TPP), which has phosphate groups. This phosphate group (negative charge) from TPP will be cross-linked to the amine group from chitosan (positive charge) (Fig. 10.2; Chávez de Paz et al., 2011; Cho et al., 2010).

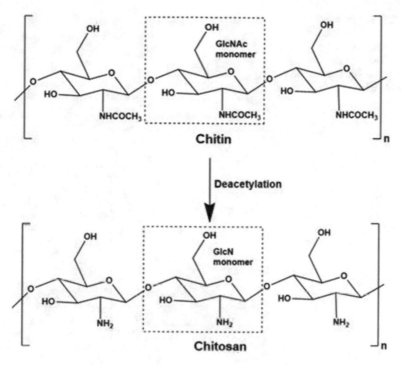

Fig. 10.1 Deacetylation of chitin to form chitosan (Schmitz et al., 2019)

The size of chitosan nanoparticles can be very small. Hence, chitosan nanoparticles can cross biological boundaries, such as tissue, thereby improving the effectiveness of the drug (Wang et al., 2011). Moreover, chitosan nanoparticles under 400 nm will be able to deliver specific drugs to cancerous tissues using a passive drug delivery system, specifically through leaky vasculature surrounding the cancer tissues (Danhier et al., 2010).

In the active drug delivery system, nanoparticles are conjugated with ligands to direct the nanoparticles to the desired cell. Folic acid is one of the ligands that is often used to designate nanoparticles in an active drug delivery system. Folic acid will bind to folic acid receptors, which are found abundantly in various types of cancer cells, such as the breast, brain, kidneys, breasts, lungs, or retinoblastoma cancer cells. Folic acid receptor expression is 100–300 times higher in cancer cells than in healthy cells (Parveen & Sahoo, 2010). Furthermore, Vllasaliu et al. (2013) reported that due to carboxyl groups in folic acid, it can be easily conjugated with other macromolecules, such as chitosan. As a potential ligand, folic acid can potentially enhance the endocytosis of chitosan nanoparticles into the targeted cells (Jin et al., 2016; Parveen & Sahoo, 2010). Yang et al. (2010) also reported that folic acid conjugated to chitosan nanoparticles could increase the accumulation of protoporphyrin IX in colorectal cancer cells. Vllasaliu et al. (2013) stated that folic acid is a

Fig. 10.2 Structure of
(**a**) chitosan, (**b**) sodium
tripolyphosphate (TPP),
(**c**) crosslink between
chitosan and TPP (Chávez de
Paz et al., 2011; Cho et al.,
2010)

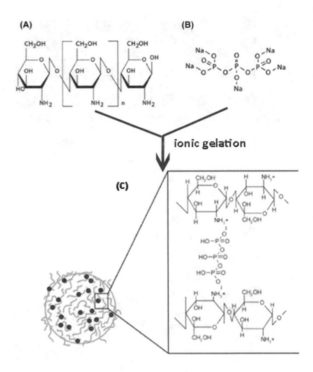

more stable and inexpensive ligand compared to other ligands, such as monoclonal antibodies.

Based on the explanation above, chitosan nanoparticles can be used to deliver a particular drug. In this research, bioactive compounds from propolis were encapsulated in chitosan nanoparticles. Propolis has antibacterial, anti-inflammatory, and anticancer activities. However, propolis has low bioavailability and is easily degraded in the body, especially in the digestive tract, so that this condition can reduce the effectiveness of propolis in the body (Elbaz et al., 2016). In order to increase the efficiency of the propolis effect, it is crucial to increase the bioavailability of propolis in the body, i.e. via folic acid conjugated chitosan nanoparticles, as a drug delivery system.

The essential parameters for nanoparticles are particle size and surface charge of particles. Generally, nanoparticle sizes must be constructed in such a way that the particles can pass through the endothelium of abnormal blood vessels in tumor tissue (Haley & Frenkel, 2008). The particle surface charge is revealed by its zeta potential value. A good zeta potential value should be more than 30 mV. The higher the zeta potential value of a particle, the greater the repulsion force between nanoparticles so that they do not quickly aggregate (Honary & Zahir, 2013).

The internalization of nanoparticles by cells occurs through various endocytic pathways. Cells can phagocytose nanoparticles bigger than 250 nm. Nanoparticles with a size of about 100 nm will be pinocytosed, whereas nanoparticles less than 100 nm will be internalized by clathrin or caveolin proteins. Generally, nanoparticles that bind to ligands from specific receptors on cancer cells will be internalized into cells by both proteins (Rajabi & Mousa, 2016).

In addition to the size of the nanoparticles, the surface charge of the nanoparticles also affects the internalization of the nanoparticles. Positively charged nanoparticles can enter the cell faster because of the strong electrostatic interaction between the nanoparticle and the negatively charged cell membrane (Rajabi & Mousa, 2016). Bannunah et al. (2014) showed that positively charged nanoparticles have been internalized better than negatively charged nanoparticles.

Synthesis of chitosan nanoparticles generally uses the ionic gelation method with polyanion compounds. The amine group in chitosan is ionized in a weak acid environment, allowing the polymer to stick to negatively charged surfaces (Loh et al., 2010). Sodium tripolyphosphate (Na-TPP) is one of the most widely used cross-linkers in the synthesis of chitosan nanoparticles because the polyanion is non-toxic and has multivalent properties (Fan et al., 2012). When the chitosan solution is mixed with the TPP solution, nanoparticles are formed by the formation of molecular bonds between phosphate in TPP and amine groups in chitosan (Fig. 10.2, Chávez de Paz et al., 2011; Cho et al., 2010).

10.3 Materials and Methods

10.3.1 Materials

For synthesizing folic acid conjugated chitosan nanoparticles containing propolis, several reagents are needed such as low molecular weight chitosan (Sigma-Aldrich), sodium tripolyphosphate (TPP; Sigma-Aldrich), folic acid, propolis. Propolis was isolated from Trigona bees (Rahmi Propolis). Other reagents used for preparing nanoparticles include EDC (1-(3-Dimethylaminopropyl)-3-ethyl carbodi-imide hydrochloride) (Sigma-Aldrich), and trehalose.

10.3.2 Extraction of Propolis

Propolis extraction was carried out to facilitate and optimize the encapsulation of bioactive propolis compounds into nanoparticles. The method used was as follows:

1. 40 mg of propolis was dissolved in 80 mL ethanol 70% (1:2) by stirring with a blender.
2. The Erlenmeyer flask containing this propolis solution was wrapped with aluminum foil, then stirred on a shaker at 150 rpm.
3. Every 24 h, the propolis solution was filtered and 80 mL of 70% ethanol was added to the propolis. This was repeated for three weeks until the propolis filtrate was clear.

4. In order to obtain a concentrated propolis extract, the propolis filtrate was evaporated with a rotary evaporator. This concentrated propolis extract contained various compounds, such as phenolic compounds.
5. The concentrated extract of propolis was then placed in an evaporating dish and dried in an oven at 40 °C until it became a propolis powder.

10.3.3 Conjugation of Folic Acid with Chitosan

Folic acid needed to be bound to chitosan before the synthesis of nanoparticles. The conjugation procedure was modified from the method developed by Yang et al. (2010), which is as follows.

1. 1.178 mg of folic acid was dissolved in 2 mL 100% DMSO. This folic acid solution was then added to 2 mL DMSO containing 1.56 mg EDC. This solution was added to 100 mL 0.5% chitosan.
2. The solution was stirred at 400 rpm for 16 h.
3. The pH of the mixture was adjusted to pH 9.0, centrifuged at 4500 rpm for 10 min, at 25 °C and the supernatant was discarded.
4. The pellet was suspended in a small volume of 2% acetic acid. The conjugated nanoparticles were purified by dialysis with a cutoff between 12,000 and 14,000 Dalton.
5. Dialysis was first carried out for one day against PBS pH 7.4, followed by another day against distilled water.
6. The conjugate was then freeze-dried.

10.3.4 Preparation of Chitosan-Folic Acid Nanoparticles

The chitosan-folic acid nanoparticles (NP-KP-P) were synthesized by modifying the method by Yang et al. (2010).

1. Chitosan-folic acid (0.05%) was dissolved in 100 mL of 2% acetic acid (pH 4.7), and stirred overnight. The chitosan-folic acid solution was then sonicated for 30 min.
2. For the process of propolis encapsulation, 0.48 mL propolis was mixed with 12 mL chitosan-folic acid solution.
3. The mixture of chitosan solution with propolis was stirred for 20 min, then 4.8 mL TPP was added dropwise to the chitosan-folic acid-propolis solution or a chitosan-folic acid solution so that nanoparticles would form.
4. The nanoparticle suspension was then centrifuged at 11,000 rpm for 1 h.
5. The pellet was then resuspended with 1 mL milli-Q water and sonicated for 30 min, while the supernatant was stored to be used later to determine the concentration of the encapsulated propolis.

6. Chitosan-folic acid (NP-KF-Blanks) and chitosan-folic acid-propolis nanoparticles (NP-KF-P) were characterized, which included the diameter, zeta potential, morphology, and chemical bonds of the nanoparticles. The morphology of the nanoparticles was examined using a scanning electron microscope. The chemical bonds in the conjugate were observed using FTIR (Fourier Transform Infrared Spectrophotometer). The diameter and zeta potential of the nanoparticles were measured using a Particle Size Analyzer (Malvern). Afterward, the nanoparticle suspension was mixed with trehalose 20% and freeze-dried. After freeze-drying, the nanoparticle suspension turned into nanoparticle powder.

10.3.5 Characterization of Nanoparticles

The encapsulation efficiency (EE) of the drug in chitosan-propolis nanoparticle (NP-KP-P) was measured using an indirect method, namely by measuring the concentration of the drug that remained in the supernatant (Bahreini et al., 2014; Xue et al., 2015). The amount of drug encapsulated in the nanoparticle was the total number of added drugs minus the amount of drug present in the supernatant (Rampino et al., 2013). The remaining drug concentration could be determined by measuring the absorbance of the supernatant at the appropriate wavelength (Xue et al., 2015) with a UV–Vis spectrophotometer. The absorbance values obtained were then compared to the standard curves of the drug. Equation (10.1) is used to calculate the efficiency of the encapsulation (EE).

$$EE(\%) = \frac{\text{Total drug concentration} - \text{residual drug concentration in the supernatant}}{\text{total drug concentration}} \times 100\%$$

(10.1)

The absorbance of NP-KF-P supernatant was measured using the Folin-Ciocalteu assay at a wavelength of 750 nm. The Folin-Ciocalteu assay was used to determine the total phenolic compounds found in propolis extract (Kubiliene et al., 2015). Supernatants from empty nanoparticles (NP-KF-Blanks) were used as blank when measuring drug-containing nanoparticle supernatants (Rampino et al., 2013).

10.3.6 In vitro Study of the Release of Propolis from Nanoparticles

This analysis was conducted by modifying the method developed by Viviek et al. (2013). A total of 5 mg of chitosan-folic acid nanoparticles containing propolis (NP-KF-P) were dissolved in 2 mL milli-Q water and placed in a dialysis tubing. The dialysis tubing containing a nanoparticle suspension was then placed in a beaker glass containing 30 mL PBS with various pH (pH 4.0; 6.0; 7.4). Propolis released from nanoparticles will pass through the dialysis membrane into the PBS solution.

The concentration of the released propolis was measured at 0, 3, 6, 9, 12, 24, 48 h and each time, the dialysis tubing was placed in new PBS. The concentration of released propolis was measured using a spectrophotometer using the Folin-Ciocalteu assay at a wavelength of 750 nm, primarily to determine the total phenolic compounds found in propolis extracts.

10.4   Result and Discussion

10.4.1   Folic Acid-Chitosan Conjugate

Carboxyl groups in folic acid would bind to the amine group in chitosan (Vllasaliu et al., 2013), which would effectively reduce the amount of free amine in chitosan. This free amine in chitosan would bind to TPP during the ionic gelation process. Hence, excessive folic acid could inhibit the formation of chitosan-folic acid nanoparticles. The result of chitosan-folic acid nanoparticle synthesis (NP-KF-blanks) and chitosan-folic acid-containing propolis nanoparticle synthesis (NP-KF-P) is presented in Fig. 10.3.

NP-KF-P was about 80–110 nm, whereas NP-KF-blanks was around 100–120 nm (Fig. 10.3). The structure of NP-KF-blanks and NP-KF-P were round, compact, and not aggregated. The zeta potential of NP-KF-blanks and NP-KF-P confirmed these results. The zeta potential of the nanoparticles was higher than 25 mV (Table 10.1), which revealed that the nanoparticles would be stable and would not easily aggregate.

TPP regulated the diameter of folic acid nanoparticles as well as the Polydispersity Index (PDI) of NP-KF-blanks. PDI is an indicator of nanoparticle distribution based on the diameter of nanoparticles in suspension. Table 10.1 shows that the PDI

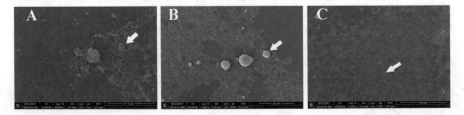

Fig. 10.3 Observation of folic acid-chitosan conjugate nanoparticle morphology using SEM. (**A**) chitosan-folic acid conjugate nanoparticles (NP-KF-blanks) (arrow); (**B**) chitosan-folic acid-containing propolis nanoparticles (NP-KF-P) (arrow); (**C**) chitosan-folic acid powder (arrow)

Table 10.1 Nanoparticle characteristics

Nanoparticle	Diameter of nanoparticle	PDI	Zeta potensial ± SD (mV)
NP-KF-blanks	129 nm	0.369	30.5 ± 1.04

of chitosan-folic acid nanoparticles (NP-KF-blanks) was 0.369. PDI (below 0.5) of NP-KF-blanks showed that the distribution of nanoparticles based on the diameter of nanoparticles was homogeneous. Rampino et al. (2013) explained that nanoparticles with PDI below 0.5 indicated that, based on the diameter of the nanoparticles, the nanoparticles were evenly distributed. Zeta potential shows the charge on the surface of nanoparticles, reflecting the stability of nanoparticles (Pan et al., 2012). Based on Table 10.1, the zeta potential of NP-KF-blanks was 30.5 mV. These results indicated that NP-KF-blanks were successfully synthesized and would not be rapidly aggregated. Jin et al. (2016) explained that a zeta potential greater than 25 mV could increase the stability of nanoparticles in nanoparticle suspensions. The positive charge of nanoparticles would repel each other so that the nanoparticles would not be aggregated (Fonte et al., 2012).

Chitosan nanoparticles (NP-K) and chitosan-folic acid nanoparticles (NP-KF-blanks) have a spherical shape with a diameter below 200 nm (Fig. 10.4). The addition of 10% ethanol containing 10% propolis (Fig. 10.4c) showed that the NP-KF was spherical, and had a diameter of less than 200 nm.

Sonication affects the formation of nanoparticles after the synthesis process. Pradhan et al. (2016) explained that sonication could distribute nanoparticles, which were likely to aggregate after synthesis. In this study, sonication was performed twice, before and after nanoparticle synthesis. Sonication of the chitosan-folic acid solution was performed to shorten the chitosan chain. The short chitosan chain would allow TPP to bind to the short-chain and form small spheres during ionic gelation. Optimization results of nanoparticle synthesis (Table 10.2) showed that without sonication, the diameter of the nanoparticles (bigger than 200 nm) was larger than that of sonication.

Fig. 10.4 Morphology of folic acid-chitosan conjugate nanoparticles using SEM. (**A**) chitosan nanoparticles (NP-K); (**B**) NP-KF-blanks; (**C**) chitosan-folic acid nanoparticles containing 10% propolis (NP-KF-P) (white arrow)

Table 10.2 Sonification effect on nanoparticle diameter

Optimasi	Type of nanopartikle	Diameter of nanoparticles
Chitosan-folic acid without sonification	NP-KF-blanks	160 nm
	NP-KF-P	239 nm
Chitosan-folic acid with sonification 30 min	NP-KF-blanks	153 nm; 129 nm
	NP-KF-P	176 nm; 153 nm

10.4.2 Encapsulation Efficiency of Propolis in Nanoparticles

The encapsulation efficiency of a drug was influenced by how many groups were successfully bound to the nanoparticles. Hydrogen bonds will be formed between functional groups in propolis and chitosan (Elbaz et al., 2016; Franca et al., 2014). Table 10.3 revealed that the encapsulation efficiency of propolis in NP-KF-P varied between 35.1 ± 2.3% and 76.4 ± 2.4%, with the concentration of propolis in the NP-KF-P between 3.87 ± 0.27 μg/mL and 8.73 ± 0.29 μg/mL.

The attainment of propolis encapsulation in nanoparticles was confirmed by the encapsulation efficiency and the NP-KF-P diameter. The results of the synthesis of chitosan-folic acid nanoparticles in this study are revealed in Table 10.4. NP-KF and NP-KF-propolis were successfully synthesized with an average diameter of 129 ± 3.4 nm and 153.9 ± 1.3 nm, respectively. The difference in diameter indicated that propolis was successfully encapsulated in chitosan-folic acid nanoparticles, thereby increasing the diameter of the nanoparticles.

The addition of propolis to chitosan-folic acid nanoparticles also did not significantly alter PDI compared to NP-KF. NP-KF-P had a PDI value of 0.380. The PDI value (< 0.5) of the NP-KF-P also showed that the diameter of the nanoparticles was relatively homogeneous. The zeta potential of the NF-KF-P was 29.7 mV (Table 10.4), indicating that NP-KF-P had been successfully synthesized and will not easily aggregate. This result was compatible with the SEM results of the NP-KF-P (Fig. 10.3b).

Table 10.3 Propolis encapsulation efficiency in NP-KF-P

Propolis encapsulation efficiency (%)	Propolis concentration in NP-KF-P (μg/mL)
76.4 ± 2.4	8.73 ± 0.29
60. 3 ± 5.3	6.93 ± 0.62
50.2 ± 0.2	5.73 ± 0,03
40.3 ± 0.6	4.6 ± 0.07
35.1 ± 2.3	3.87 ± 0.27

Table 10.4 Diameter, distribution and zeta potensial of chitosan-folic acid nanoparticles

Nanoparticle	Diameter of nanoparticle ± SD (nm)	PDI	Zeta potensial ± SD (mV)
Chitosan-folic acid nanoparticle blanks (NP-KF-blanks)	129 ± 3.4	0.369	30.5 ± 1.04
Chitosan-folic acid-propolis nanoparticles (NP-KF-P)	153.9 ± 1.3	0.380	29.7 ± 0.82

PDI = Polydispersity Index

10.4.3   FTIR Analysis

FTIR analysis was used to distinguish among folic acid conjugated chitosan, folic acid conjugated chitosan nanoparticles, and folic acid chitosan nanoparticles containing propolis. Figure 10.4 showed that chitosan had a broad absorption area for O–H and N–H groups. The conjugation of chitosan with folic acid narrowed the absorption area of the O–H and N–H (Fig. 10.2b-1), which was probably caused by the appearance of a suspected bond between the N–H group in chitosan with the carboxyl group in folic acid. Additionally, there was an amide group [C(=O)N] at the wavenumber of 1656.85 cm^{-1}, which was a unique bond between the amine (in chitosan) and the carboxyl group (in folic acid). These FTIR results showed that chitosan was conjugated with folic acid.

The FTIR analysis on chitosan-folic acid blank (NP-KF-blanks) nanoparticles showed that the absorption area of O–H and N–H groups of this nanoparticle was smaller than (Fig. 10.5c-1) chitosan-folic acid powder (10.5b-1). This feature was

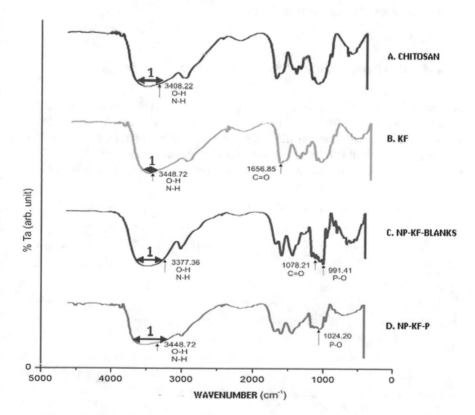

Fig. 10.5 FTIR results of chitosan, KF, NP-KF-blanks, NP-KF-P. KF = chitosan folic acid; NP-KF = chitosan-folic acid nanoparticles; NP-KF-P = NP-KF containing propolis. (1) broad absorption area of O–H and N–H groups

probably affected by a bond between the N–H group from chitosan and the phosphate group from TPP. According to Mattu et al. (2013, the phosphate group in TPP would bind to the amine group in chitosan), in the formation of nanoparticles. The FTIR result of NP-KF-P also showed the presence of phosphate groups (P-O) from TPP at a wavenumber of 991.41. According to Pramanik et al. (2009), the phosphate group (P-O) from TPP bound to chitosan absorbed the infrared light in the wavenumber range 493–1047.

FTIR of chitosan-folic acid nanoparticles containing propolis (NP-KF-P) showed that propolis had been encapsulated in chitosan-folic acid nanoparticles. The addition of propolis in the chitosan-folic acid nanoparticles broadens the absorption area of the O–H and N–H groups (Fig. 10.5d-1) compared to the chitosan-folic acid nanoparticles blanks (Fig. 10.5c-1). The O–H group of propolis might affect the widening of the O–H and N–H groups' absorption area in chitosan-folic acid nanoparticles. Huang et al. (2014) stated that propolis contains various flavonoid compounds that have O–H groups. Hasan et al. (2013) also explained that the FTIR analysis of propolis compounds showed a broad O–H group at a wavenumber of 3267 due to its phenolic compounds, especially flavonoids.

10.4.4 In vitro *Study of Propolis Released from NP-KF-P*

As shown in Fig. 10.6, the release of propolis from NP-KF-P occurred quickly in an acidic environment (pH 4 and 6). After 48 h in an acidic environment, 89% of propolis was released from NP-KF-P at pH 4 and 60% was released at pH 6. In contrast, only a small percentage of propolis (13%) was released from NP-KF-P at pH 7. This condition would be beneficial for the propolis delivery system because the

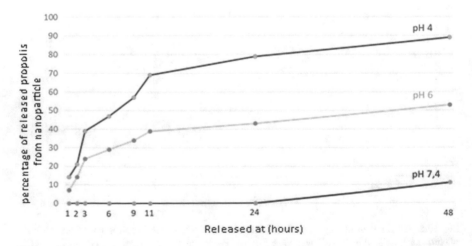

Fig. 10.6 In vitro study of propolis released from NP-KF-P at three different pH variations

nanoparticle could be delivered into the cells, and then inside the cell, propolis could be released inside the lysosome, which has an acidic environment. Alternatively, the NP-KF-P could be transported safely through the blood (pH 7) to the tumor tissue. The propolis would be released around the tumor tissue due to its acid environment.

10.5   Conclusion

The synthesis of folic acid conjugated chitosan nanoparticles that encapsulated propolis is regulated by several factors, including folic acid:chitosan ratio, the molecular weight of chitosan, the proportion of chitosan-folic acid to TPP during ionic gelation, the concentration of propolis used, and the duration of sonication. Spherical propolis conjugated chitosan folic acid nanoparticles were successfully synthesized with nanoparticles with a diameter of 129 ± 3.4 nm (NP-KF-blanks) and 153.9 ± 1.3 nm (NP-KF-P). Sonication of the chitosan-folic acid solution before synthesis for 30 min reduced the diameter of chitosan-folic acid-propolis nanoparticles, with a propolis encapsulation efficiency between $35.1 \pm 2.3\%$ and $76.4 \pm 2.4\%$. Due to the solubility of chitosan in an acidic environment, the percentage of drug released was highest at pH 4 (about 90%).

References

Bannunah, A. M., Vllasaliu, D., Lord, J., & Stolnik, S. (2014). Mechanisms of nanoparticle internalization and transport across an intestinal epithelial cell model: Effect of size and surface charge. *Mol Pharmaceutics, 11*(12), 4363–4373.

Bahreini, E., Aghaiypour, K., Abbasalipourkabir, R., Mokarram, A. R., Goodarzi, M. T., & Saidijam, M. (2014). Preparation and nanoencapsulation of l-asparaginase II in chitosan-tripolyphosphate nanoparticles and in vitro release study. *Nanoscale Research Letters, 9*(1), 340.

Chávez de Paz L. E., Resin, A., Howard, K. A., Sutherland, D. S., & Wejse P. L. (2011). Antimicrobial effect of chitosan nanoparticles on streptococcus mutans biofilms. *Applied and Environmental Microbiology, 77*(11): 3892–3895.

Cheng, L., Ma, H., Shao, M., Fan, Q., Lv, H., Peng, J., Hao, T., Li, D., Zhao, C., & Zong, X. (2017). Synthesis of folate chitosan nanoparticles loaded with ligustrazine to target folate receptor positive cancer cells. *Molecular Medicine Reports, 16*(2), 1101–1108.

Cho, Y., Shi, R., & Borgens, R. B. (2010). Chitosan nanoparticle-based neuronal membrane sealing and neuroprotection following acrolein-induced cell injury. *Journal of Biological Engineering, 4*, 2.

Danhier, F., Feron, O., & Préat, V. (2010). To exploit the tumor microenvironment: Passive and active tumor targeting of nanocarriers for anti-cancer drug delivery. *Journal of Controlled Release, 148*, 135–146.

Elbaz, N. M., Khalil, I. A., Abd-Rabou, A. A., & El-Sherbiny, I. M. (2016). Chitosan-based nano-in-microparticle carriers for enhanced oral delivery and anticancer activity of propolis. *International Journal of Biological Macromolecules, 92*, 254–269.

Fan, W., Yan, W., Xu, Z., & Ni, H. (2012). Formation mechanism of monodisperse, low molecular weight chitosan nanoparticles by ionic gelation technique. *Colloids and Surfaces B: Biointerfaces, 90*, 21–27.

Fonte, P., Andrade, F., Araújo, F., Andrade, C., das Neves, J., & Sarmento, B. (2012). Chitosan-coated solid lipid nanoparticles for insulin delivery. *Methods Enzymol, 508*, 295–314.

Franca, J. R., De Luca, M. P., Ribeiro, T. G., Castilho, R. O., Moreira, A. N., Santos, V. R., & Faraco, A. A. (2014). Propolis-based chitosan varnish: Drug delivery, controlled release and antimicrobial activity against oral pathogen bacteria. *BMC Complementary and Alternative Medicine, 14*(1), 478.

Hasan, A. E. Z., Mangunwidjaja, D., Sunarti, T. J., Suparno, O., & Setiyono, A. (2013). Optimasi ekstraksi propolis menggunakan cara maserasi dengan pelarut etanol 70% dan pemanasan gelom-bang mikro serta karakterisasinya sebagai bahan antikanker payudara. *Jurnal Teknologi Industri Pertanian, 23*(1), 13–21.

Haley, B., & Frenkel, E. (2008). Nanoparticles for drug delivery in cancer treatment. *Urologic Oncology, 26*(1), 57–64.

Honary, S., & Zahir, F. (2013). Effect of zeta potential on the properties of nano-drug delivery systems—A Review (Part 1). *Tropical Journal of Pharmaceutical Research, 12*(2).

Huang, S., Zhang, C. P., Wang, K., Li, G. Q., & Hu, F. L. (2014). Recent advances in the chemical composition of propolis. *Molecules, 19*(12), 19610–19632.

Jin, H., Pi, J., Yang, F., Jiang, J., Wang, X., Bai, H., Shao, M., Huang, L., Zhu, H., Yang, P., & Li, L. (2016). Folate-chitosan nanoparticles loaded with ursolic acid confer anti-breast cancer activities in vitro and in vivo. *Science and Reports, 6*, 30782.

Kean, T., & Thanou, M. (2010). Biodegradation, biodistribution and toxicity of chitosan. *Advanced Drug Delivery Reviews, 62*, 3–11.

Kubiliene, L., Laugaliene, V., Pavilonis, A., Maruska, A., Majiene, D., Barcauskaite, K., Kubilius, R., Kasparaviciene, G., & Savickas A. (2015). Alternative preparation of propolis extracts: Comparison of their composition and biological activities. *BMC Complementary and Alternative Medicine, 15*.

Loh, J. W., Yeoh, G., Saunders, M., & Lim, L. Y. (2010). Uptake and cytotoxicity of chitosan nanoparticles in human liver cells. *Toxicology and Applied Pharmacology, 249*(2), 148–157.

Mattu, C., Li, R., & Ciardelli, G. (2013). Chitosan nanoparticles as therapeutic protein nanocarriers: The Effect of pH on Particle Formation and Encapsulation Efficiency. *Polymer Composites, 34*(9), 1538–1545.

Mohanraj, V. J., & Chen, Y. (2006). Nanoparticles-a review . *Tropical Journal of Pharmaceutical Research, 5*(1), 561–573.

Pan, H., Marsh, J. N., Christenson, E. T., Soman, N. R., Ivashyna, O., Lanza, G. M., Schlesinger, P. H., & Wickline, S. A. (2012). Postformulation peptide drug loading of nanostructures. *Methods in Enzymology, 508*, 17–39.

Parveen, S., & Sahoo, S. K. (2010). Evaluation of cytotoxicity and mechanism of apoptosis of doxorubicin using folate-decorated chitosan nanoparticles for targeted delivery to retinoblastoma. *Cancer Nanotechnology, 1*(1), 47.

Pradhan, S., Hedberg, J., Blomberg, E., Wold, S., & Wallinder, I. O. (2016). Effect of sonication on particle dispersion, administered dose and metal release of non-functionalized, non-inert metal nanoparticles. *Journal of Nanoparticle Research, 18*(9), 285.

Pramanik, N., Mishra, D., Banerjee, I., Maiti, T. K., Bhargava, P., & Pramanik, P. (2009). Chemical synthesis, characterization, and biocompatibility study of hydroxyapatite/chitosan phosphate nanocomposite for bone tissue engineering applications. *International Journal of Biomaterials.* https://doi.org/10.1155/2009/512417.

Rampino, A., Borgogna, M., Blasi, P., Bellich, B., & Cesàro, A. (2013). Chitosan nanoparticles: preparation, size evolution and stability. *International Journal of Pharmaceutics, 455*(1–2), 219–228.

Rajabi, M., & Mousa, S. (2016). Lipid nanoparticles and their application in nanomedicine. *Current Pharmaceutical Biotechnology, 17*, 662–672.

Riva, R., Ragelle, H., Rieux, A. D., Duhem, N., Jérôme, C., & Préat, V. (2011). Chitosan and chitosan derivatives in drug delivery and tissue engineering. *Advances in Polymer Science, 244*, 19–44.

Schmitz, C., Auza, L. G., Koberidze, D., Rasche, S., Fischer, R., & Bortesi, L. (2019). Conversion of chitin to defined chitosan oligomers: Current status and future prospects. *Marine Drugs, 17*(8), 452.

Vllasaliu, D., Casettari, L., Bonacucina, G., Cespi, M., Palmieri, G. F., & Illum, L. (2013). Folic acid conjugated chitosan nanoparticles for tumor targeting of therapeutic and imaging agents. *Pharm Nanotechnol, 1*, 184.

Viviek, R., Babu, V. N., Thangam, R., Subramanian, K. S., & Kannan, S. (2013). pH-responsive drug delivery of chitosan nanoparticles as tamoxifen carriers for effective anti-tumor activity in breast cancer cells. *Colloids and Surfaces B, 111*, 117–123.

Wang, J. J., Zeng, Z. W., Xiao, R. Z., Xie, T., Zhou, G. L., Zhan, X. R., & Wang, S. L. (2011). Recent advances of chitosan nanoparticles as drug carriers. *International Journal of Nanomedicine, 6*, 765–774.

Wang, F., Wang, Y., Ma, Q., Cao, Y., & Yu, B. (2017). Development and characterization of folic acid-conjugated chitosan nanoparticles for targeted and controlled delivery of gemcitabinein lung cancer therapeutics. *Artificial Cells, Nanomedicine, and Biotechnology, 45*(8), 1530–1538.

Wilczewska, A., Niemirowicz, K., Markiewicz, K. H., & Car, H. (2012). Nanoparticles as drug delivery systems. *Pharmacological Reports, 64*(5), 1020–1037.

Xue, M., Hu, S., Lu, Y., Zhang, Y., Jiang, X., An, S., Guo, Y., Zhou, X., Hou, H., & Jiang, C. (2015). Development of chitosan nanoparticles as drug delivery system for a prototype capsid inhibitor. *International Journal of Pharmaceutics, 495*(2).

Yang, S. J., Lin, F. H., Tsai, K. C., Wei, M. F., Tsai, H. M., Wong, J. M., & Shieh, M. J. (2010). Folic acid-conjugated chitosan nanoparticles enhanced protoporphyrin ix accumulation in colorectal cancer cells. *Bioconjugate Chemistry, 21*(4), 679–689.

Zhang, J., Lan, C. Q., Post, M., Simard, B., Deslandes, Y., & Hsieh, T. H. (2006). Design of nanoparticles as drug carriers for cancer therapy. *Cancer Genomics and Proteomics, 3*, 147–157.

Printed in the United States
by Baker & Taylor Publisher Services